T0142955

Supervised Descriptive Pattern Mining

Sebastián Ventura • José María Luna

Supervised Descriptive Pattern Mining

Springer

Sebastián Ventura
Computer Science
University of Cordoba
Cordoba, Spain

José María Luna
Computer Science
University of Cordoba
Cordoba, Spain

ISBN 978-3-030-07456-2   ISBN 978-3-319-98140-6 (eBook)
https://doi.org/10.1007/978-3-319-98140-6

This Springer imprint is published by the registered company Springer Nature Switzerland AG
The registered company address is: Gewerbestrasse 11, 6330 Cham, Switzerland

To Marta and Laura, my favorite noise makers.

S. Ventura

To my son Luca, the one who makes me smile every day.

J. M. Luna

Acknowledgments

The authors would like to thank the Springer editorial team for giving them the opportunity to publish this book, for their great support toward the preparation and completion of this work, and their valuable editing suggestions to improve the organization and readability of the manuscript. The authors also want to thank anyone who helped them during the preparation of the book, whose comments were very helpful for improving its quality.

The authors would also like to thank the Spanish Ministry of Economy and the European Fund for Regional Development for supporting their research toward the project TIN2017-83445-P.

Contents

Chapter 1
Introduction to Supervised Descriptive Pattern Mining

Abstract This chapter introduces the supervised descriptive pattern mining task to the reader, providing him/her with the concept of patterns as well as presenting a description of the type of patterns usually found in literature. Patterns on advanced data types are also defined, denoting the usefulness of sequential and spatiotemporal patterns, patterns on graphs, high utility patterns, uncertain patterns, along with patterns defined on multiple-instance domains. The utility of the supervised descriptive pattern mining task is analysed and its main subtasks are formally described, including contrast sets, emerging patterns, subgroup discovery, class association rules, exceptional models, among others. Finally, the importance of analysing the computational complexity in the pattern mining field is also considered, examining different ways of reducing this complexity.

1.1 Patterns in Data Analysis

In many application fields, there is a real incentive to collect, manage and transform raw data into significant and meaningful information that may be used for a subsequent analysis that leads better decision making [79]. In these fields there is therefore a growing interest in data analysis, which is concerned with the development of methods and techniques for making sense of data [31]. When talking about data analysis, the key element is the pattern, representing any type of homogeneity and regularity in data, and serving as a good descriptor of intrinsic and important properties of data [3]. A pattern is a set of elements (items) that are related in a database. Formally speaking, a pattern P in a database Ω is defined as a subset of items $I = \{i_1, ..., i_n\} \in \Omega$, i. e. $P \subseteq I$, that describes valuable features of data [83]. Additionally, the length or size of a pattern P is denoted as the number of single items that it comprises, and those patterns consisting of a unique item are called singletons.

Pattern mining is a really alluring task whose aim is to find all patterns that are present in at least a fraction of the transactions (records) in a database. This task is being applied to more and more application fields nowadays, market basket analysis [12] being the first application domain in which it was correctly applied. In

© Springer Nature Switzerland AG 2018

S. Ventura, J. M. Luna, *Supervised Descriptive Pattern Mining*,
https://doi.org/10.1007/978-3-319-98140-6_1

this very field, a high number of purchases are usually bought on impulse, requiring an in-depth analysis to obtain clues that determine which specific items tend to be strongly related [9]. As a matter of example, thanks to pattern mining, it is possible to determine how likely is for a customer to buy beer and diapers together, so the interest of these items might be quantified by their probability of co-occurrence. This analysis might allow shopkeepers to increase the sales by re-locating the products on the shelves, or even it might allow managers to plan diverse advertising strategies.

The aim of any pattern mining algorithm is to make sense of any mess in data, arranging elements of data in order to obtain those sets that most frequently appear or those with a special interest for a specific aim. Considering the *Jenga* game (see Fig. 1.1), the aim is to extract any blocks (patterns) that share information from a tower constructed of m blocks (dataset). The mining of patterns requires high efficient algorithms which computational cost highly depends on the dimensionality of the dataset to be analysed. Given a dataset comprising n singletons or single items, i.e. $I = \{i_1, i_2, ..., i_n\}$, a maximum of $2^n - 1$ different patterns can be found, so any straightforward approach becomes extremely complex with the increasing number of singletons. This challenging task is often simplified by pruning the search space to those patterns of interest according to some constraints, reducing the time and resource requirements. The most-well known pruning strategy in pattern mining is based on the anti-monotone property, which denotes that if a pattern P does not frequently appear in data then none of its supersets are frequent. This interesting property gives rise to a reduction in the search space since once an infrequent pattern is found, none of its supersets need to be generated/analysed.

According to the previous description, each pattern represents a data subset and its frequency is related to the size of such subset. In this regard, local

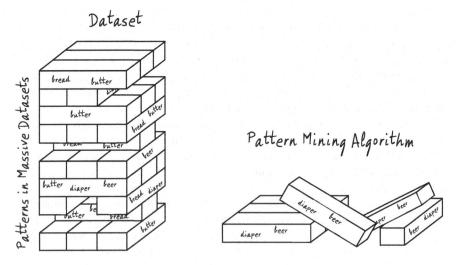

Fig. 1.1 Mining patterns in massive datasets by arranging elements

patterns represent small parts of data, with a specific internal structure, and which distribution substantially deviate from the distribution of the whole data. Local patterns (less frequent) are therefore more relevant than global patterns (very frequent). According to some authors [25], focusing on local patterns is more appropriate than on global patterns since the latter can be discovered by humans and most likely have already been discovered—there is no need to rediscover this knowledge by pattern mining approaches. Sometimes, either local or global patterns are used to describe important properties of two or more data subsets previously identified or labeled, transforming therefore the pattern mining concept into a more specific one, *supervised descriptive pattern mining* [66].

1.2   Pattern Mining: Types of Patterns and Advanced Data Types

Pattern mining was originally introduced in the early nineties in the context of market basket data with the aim of finding frequent groups of items that were bought together [4]. The problem of mining patterns is that of finding sets of items that are present in at least a fraction of the records. Thus, these patterns represent any type of homogeneity and regularity in data, and they serve as good descriptors of intrinsic and important properties of data. As a result, diverse types of patterns have been defined, which have been used as representative elements of data on different application fields, including medicine [67], education [72], and fraud detection [74] among others.

1.2.1   Frequent/Infrequent Patterns

Let $I = \{i_1, i_2, ..., i_n\}$ be the set of n items included in a dataset Ω, and let $T = \{t_1, t_2, ..., t_m\}$ be the collection of transactions such as $\forall t \in T : t \subseteq I \in \Omega$. Additionally, let P be a pattern comprising a set of items, i.e. $P \subseteq I$. The frequency (also known as support) of the pattern P is defined as the percentage of transactions in which $P \subseteq t : t \in T$ appears. It is also described in absolute terms as the number of different transactions in which P is included, i.e. $|\{\forall t \in T : P \subseteq t, t \subseteq I \in \Omega\}|$. Hence, the problem of mining frequent (or infrequent) patterns is the one of extracting those patterns that are regularly (or rarely) present in a dataset.

The pattern mining problem, considering either frequent or infrequent itemsets, can be easily described from a sample dataset (see Table 1.1) based on ten different transactions containing items (products) purchased by customers. In this example, and considering the number of patterns of size k as $\binom{n}{k} = \frac{n!}{k!(n-k)!}$ for any $k \leq n$, there are $\binom{4}{1} = 4$ patterns including a single item; $\binom{4}{2} = 6$ patterns comprising 2 items; $\binom{4}{3} = 4$ patterns containing 3 items; and $\binom{4}{4} = 1$ pattern including all

Table 1.1 Sample market basket dataset

Items
{Bread}
{Bread, Beer}
{Bread, Beer, Diaper}
{Bread, Butter}
{Bread, Beer, Butter, Diaper}
{Beer}
{Beer, Diaper}
{Bread, Butter}
{Bread, Beer, Diaper}
{Bread}

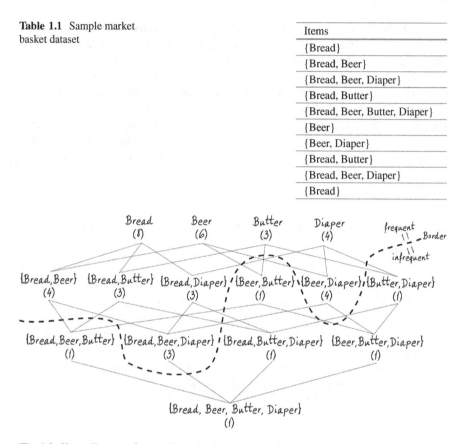

Fig. 1.2 Hasse diagram of a sample market basket dataset including ten different customers

the items. Thus, the number of feasible patterns is formally calculated as $2^n - 1$ for any dataset comprising n singletons (see the Hasse diagram shown in Fig. 1.2 where the number of transactions in which each pattern appears is illustrated into brackets). As a result, the complexity of finding patterns of interest is in exponential order, and this complexity is even higher when the frequency of each pattern is calculated [33], resulting as $O((2^n - 1) \times m \times n)$. Continuing with the same example, and considering a minimum support threshold of 2 (in absolute terms) in this very example, a frequent pattern is anyone with a support value greater or equal to 2. According to the Hasse diagram shown in Fig. 1.2 any pattern above the dashed line is a frequent pattern, whereas those below the dashed line are infrequent patterns.

The task of mining frequent patterns is an intensively researched problem in data mining [31] and numerous algorithms have been proposed either from exhaustive search [3] and heuristic [83] points of view. Focusing on exhaustive search approaches, Apriori [4] is a *de facto* algorithm for mining frequent patterns, and it is based on a level-wise paradigm in which all the frequent patterns of length

Fig. 1.3 Prefix-tree structure obtained by using the sample market basket dataset shown in Table 1.1

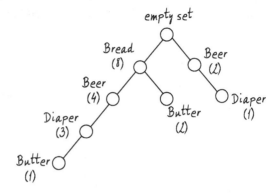

$n + 1$ are generated by using all the frequent patterns of length n. Apriori is based on the fact that the frequency of a pattern P is always lower or equal to the frequency of any of its subsets. It implies that if a pattern $P = \{i_1, i_2\}$ is frequent, both i_1 and i_2 should also be frequent. Another well-known approach for mining frequent patterns, called ECLAT, was proposed by *Zaki* [92] where the original dataset is transformed into a vertical database. Each single item is stored in the dataset together with a list of transaction-IDs where the item can be found—*Bread* in Table 1.1 is stored as $t(Bread) = \{t_1, t_2, t_3, t_4, t_5, t_8, t_9, t_{10}\}$—and the frequency of a pattern P is computed as the length of the transaction-IDs list. ECLAT obtains new patterns by using recursive intersections of previously found patterns. Finally, it is important to describe one of the most efficient exhaustive search approaches in pattern mining, that is, the FP-Growth algorithm [32], which reduces the database scans by considering a compressed representation of the database. This representation is based on a prefix-tree structure that depicts each node by using a singleton and the frequency of a pattern is denoted by the path from the root to that node. Singletons are first ranked according to their frequency so they will appear in the tree in this order. As a matter of example, and considering the aforementioned sample market basket dataset (see Table 1.1), the FP-Growth algorithm analyses each transaction in order to construct the tree structure (see Fig. 1.3). A major feature of this algorithm is the fact that once the prefix-tree structure is obtained, then no further passes over the dataset are required and any frequent pattern can be extracted by exploring the tree from the bottom up.

1.2.2 Positive/Negative Patterns

The problem of mining patterns is generally described as the one of mining sets of items that frequently (or infrequently) appear in data. In some specific domains [47], however, it is also possible to look for those elements that are not present in the data records. In this regard, a pattern is defined as positive if all of its items are present in data records, denoting a co-occurrence among elements that appear in a set of data records. On the contrary, a pattern is negative if it describes a co-occurrence

Table 1.2 Binary representation of the sample market basket dataset (Table 1.1) considering negative items

T	Bread	$\neg$Bread	Beer	$\neg$Beer	Butter	$\neg$Butter	Diaper	$\neg$Diaper
t_1	1	0	0	1	0	1	0	1
t_2	1	0	1	0	0	1	0	1
t_3	1	0	1	0	0	1	1	0
t_4	1	0	0	1	1	0	0	1
t_5	1	0	1	0	1	0	1	0
t_6	0	1	1	0	0	1	0	1
t_7	0	1	1	0	0	1	1	0
t_8	1	0	0	1	1	0	0	1
t_9	1	0	1	0	0	1	1	0
t_{10}	1	0	0	1	0	1	0	1

between items that may be present or absent in data. For a better understanding, let us considered, in a binary format, the sample market basket dataset illustrated in Table 1.1. Now, each item is represented in both positive and negative form (see Table 1.2) and the number of singletons is therefore double (each single item is also described in its negative form). Hence, the search space highly increases since there are $\binom{8}{1} = 8$ patterns including a single item; $\binom{8}{2} = 28$ patterns comprising 2 items; $\binom{8}{3} = 56$ patterns containing 3 items; $\binom{8}{4} = 70$ patterns containing 4 items; $\binom{8}{5} = 56$ patterns containing 5 items; $\binom{8}{6} = 28$ patterns containing 6 items; $\binom{8}{7} = 8$ patterns containing 7 items; and $\binom{8}{8} = 1$ pattern including all the items. The resulting search space includes $2^8 - 1 = 255$ feasible patterns, which is larger than the one that comprises only positive patterns (15 feasible patterns can be found). Finally, it should be highlighted that the process of mining negative frequent patterns can be similarly carried out to the one of mining positive frequent patterns [93]. The only requirement is that negative elements should be added to the dataset (see Table 1.2) so the mining process may follow the same process as if all the singletons were positive.

Different properties can be described for mining positive/negative patterns. First, items related to their opposite forms, for example $\{Bread, \neg Bread\}$, always produce a zero support value since records cannot satisfy both at time, i.e. $\forall i \in \Omega : t(i) \cap t(\neg i) = \emptyset$. Second, the frequency of a singleton that is represented in its negative form is equal to the number of transactions (data records) minus the frequency of the same item in its positive form, and viceverse. For example, $f(Bread) = |T| - f(\neg Bread) = 8$. Third, all the combinations of positive/negative items in a pattern will have an absolute support equal to the number of transactions in data, e.g. $f(Bread) + f(\neg Bread) = |T|$. This assertion is true for patterns of any length (number of items it includes) so considering the elements $Bread$ and $Diaper$, then it is satisfied that $f(\{Bread, Diaper\}) + f(\{Bread, \neg Diaper\}) + f(\{\neg Bread, Diaper\}) + f(\{\neg Bread, \neg Diaper\}) = |T|$.

1.2.3   Maximal/Closed/Colossal Patterns

Since pattern mining is able to extract any feasible combination of elements in data, a high computational time is required when the number of different elements is too high. According to the aforementioned property used by Apriori to discard infrequent patterns, it is possible to assert that a pattern P is always lower or equal to the frequency of any of its subsets, which also means that if P is frequent, then any of its subsets are also frequent. Based on this issue and considering a lengthy frequent pattern, a significant amount of memory and time can be spent on counting redundant patterns [3]. For instance, given a pattern P of length $|P| = 10$, it comprises a total of $2^{10} - 2 = 1022$ sub-patterns that are also frequent. In this sense, condensed representations of frequent patterns were proposed as a solution to overcome the computational and storage problems. These compact representations not only reduce the computational time and the memory requirements but also ease the knowledge discovery process carried out after the mining.

In a formal way, and considering all the frequent patterns $\mathscr{P} = \{P_1, P_2, ..., P_m\}$ found in a dataset Ω, a frequent pattern $P_i \in \mathscr{P}$ is defined as maximal frequent pattern [83] if and only if it has no frequent superset, i.e. $/\exists P_j \supset P_i :$ $P_i, P_j \in \mathscr{P}$. As a result, the number of maximal frequent patterns is considerably smaller than the number of all the frequent patterns. As a matter of clarification, let us consider the market basket dataset shown in Table 1.1 where the set of frequent patterns $\mathscr{P}$ was illustrated in Fig. 1.2 (patterns above the dashed line). Based on the aforementioned example, Fig. 1.4 shows all the existing maximal frequent patterns for a minimum absolute support threshold value of 2. The pattern $P_i = \{Bread, Beer, Diaper\}$ is a maximal frequent pattern since its superset $\{Bread, Beer, Butter, Diaper\}$ is not frequent. Similarly, the pattern

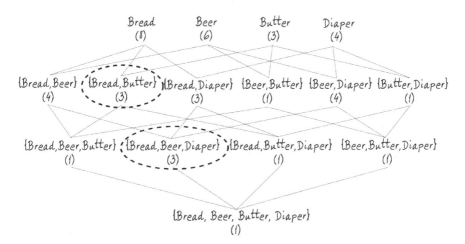

Fig. 1.4 Maximal frequent patterns represented in a Hasse diagram of a sample market basket dataset

$P_j = \{Bread, Butter\}$ is a maximal frequent pattern since none of its supersets, i.e. $\{Bread, Beer, Butter\}$ and $\{Bread, Butter, Diaper\}$, are frequent.

Condensed representations of frequent patterns can also be carried out by means of closed frequent patterns [83], which are formally defined as those frequent patterns belonging to $\mathscr{P}$ that have no frequent superset with their same frequency of occurrence, i.e. $\nexists P_j \supset P_i : f(P_j) = f(P_i) \wedge P_i, P_j \in \mathscr{P}$. For a better understanding, let assume the same market basket dataset shown in Table 1.1 as well as its representation in a Hasse diagram (see Fig. 1.5). Here, and considering as frequent patterns those whose absolute support value is greater than 2, $\{Bread\}$ is a frequent closed pattern since it has no frequent superset with its same support value. Additional frequent closed patterns found in this dataset are: $\{Beer\}$, $\{Bread, Beer\}$, $\{Bread, Butter\}$, $\{Beer, Diaper\}$, and $\{Bread, Beer, Diaper\}$.

At this point, it is interesting to remark that maximal frequent patterns are also closed frequent patterns since they have no frequent superset and, therefore, none of their supersets have their same support. The main difference between these two types of frequent patterns lies in the fact that maximal frequent patterns may cause a loss of information (the exact frequencies of their subsets cannot be directly derived). On the contrary, closed frequent patterns do not lose information since the frequencies of their subsets can be easily derived.

The introduction of closed and maximal frequent patterns [3] partially alleviated the redundancy problem caused by lengthy patterns that include a large number of frequent subsets (a total of 1022 subpatterns can be derived from a pattern of length 10 as it was stated above). In many application domains [60], on the contrary, extremely large patterns are desired, which are denoted as colossal patterns [95]. A colossal pattern P is formally defined as a frequent pattern whose length $|P|$ is too high, and any of its sub-patterns has a similar frequency. Similarly, a core pattern

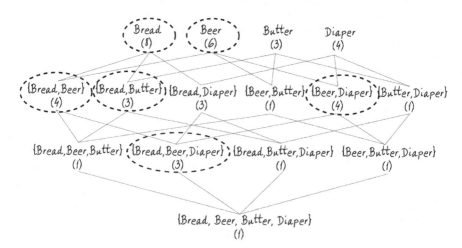

Fig. 1.5 Closed frequent patterns represented in a Hasse diagram of a sample market basket dataset

P' is a sub-pattern of a colossal pattern P having a similar frequency, i.e. $P' \subset P$, and $f(P') \approx f(P)$. This similar frequency is defined by means of the core ratio τ, which determines that $f(P)/f(P') = \tau, 0 < \tau \leq 1$. As a result, any colossal pattern tends to have a high number of core patterns.

1.2.4 Condensed Patterns

Frequent patterns [3] are almost the most widely-studied category of patterns and many efficient approaches have been proposed to extract any feasible frequent pattern in data. To date, it is still expensive to find the complete set of frequent patterns, since a dataset containing only a single transaction with 100 different elements or singletons will require to compute $2^{100} - 1 \approx 1.27 \times 10^{30}$ frequent patterns. In this regard, this daunting problem can be eased by computing a small subset of frequent patterns, named condensed patterns [71], and then use it to approximate the supports of arbitrary frequent patterns.

In some application domains [82], however, it is enough to generate patterns with a sufficiently good approximation (an estimated value with a guaranteed maximal error bound) of the support value instead of in full precision. This way of dealing with patterns is sometimes preferable due to the following reasons: (1) when dealing with really large datasets, small deviations often have minor effects on the analysis. Thus, the fact that a discovered pattern $P = \{Beer, Diaper\}$ appears 100,000 times in a dataset containing 100 million transactions provides the same information than asserting that the same pattern appears $100,000 \pm 50$ times. (2) Computing condensed patterns leads to more efficient approaches since it is only required to operate with and access to a small portion of frequent patterns [77].

For the sake of approximating the supports a function $\mathscr{F}$ is considered, which is applied to the set of unknown patterns based on the set of condensed patterns. As a matter of example, and considering the set of items $I = \{i_1, i_2, i_3, i_4\}$ in a dataset Ω, the following condensed patterns may be produced by a condensed pattern mining approach (absolute support values are denoted in brackets): $\{i_1\}(5)$, $\{i_2\}(5)$, $\{i_3\}(4)$, $\{i_4\}(4)$, $\{i_1, i_3, i_4\}(1)$, $\{i_1, i_2, i_3, i_4\}(1)$. Additionally, the following approximation function is also provided for any pattern P: $\mathscr{F}(P) = 0$ if there exists no superset $P' \supseteq P$ such as P' is defined as a condensed pattern; whereas $\mathscr{F}(P) = [support(P') - 2, support(P')]$ being $support(P')$ the minimum support for any $P' \subset P$ such as P' is defined as a condensed pattern. Thus, considering the aforementioned set of condensed patterns, it is possible to approximate the support of the pattern $P' = \{i_1, i_3\}$, which is calculated as $\mathscr{F}(P') = [4 - 2, 4] = [2, 4]$. Similarly, the support of the pattern $P'' = \{i_1, i_2\}$ can be approximated as $\mathscr{F}(P'') = [5 - 2, 5] = [3, 5]$.

1.2.5   Patterns on Advanced Data Types

Patterns are generally defined on tabular sets where each row represents a different transaction or data record. Sometimes, though, datasets are structured as a binary table where each row states for a data record and each column defines a different item that may or may not appear in the record. Nevertheless, with the increasing interest in data storage, data representation has been extended to various advanced data types including temporal data, spatiotemporal data, graphs, and high utility data among others. All the existing data types used in mining patterns are briefly studied below.

1.2.5.1   Data Streams

Generally, datasets are static sets of data represented in different ways such as unstructured sets (text documents), semistructured (XML documents), or highly structured (graphs). Sometimes, however, datasets are defined in a streaming environment [26]. A streaming dataset Ω is a sequence of transactions of indefinite length occurring at a time T_j, that is $\Omega = \{T_1, T_2, ..., T_n\}$. At each time T_j, a data window $T_{i,j}$ is defined to include the sequence of transactions occurring from time T_i to the present T_j, that is $T_{i,j} = \{T_i, T_{i+1}, ..., T_j\}$. Thus, pattern mining on data streams is similar to the task on static sets of data [3] except for the sliding window that makes the data updated and so are the results.

As an example, let us consider the sample market basket dataset shown in Table 1.1 where each data record represents a different time T_j. Here, $T_1 = \{Bread\}$, $T_2 = \{Bread, Beer\}$, $T_3 = \{Bread, Beer, Diaper\}$, etc. For a period $T_{1,5}$, the following patterns can be obtained (the frequency of each pattern is denoted into brackets): $\{Bread\}(5)$ is the only pattern (singleton) that appears 5 times in $T_{1,5}$; some other patterns $\{Beer\}(3)$ and $\{Bread, Beer\}(3)$ appear 3 times in $T_{1,5}$; other patterns such as $\{Butter\}(2)$, $\{Diaper\}(2)$, $\{Bread, Diaper\}(2)$, $\{Beer, Diaper\}(2)$, $\{Bread, Butter\}(2)$, and $\{Bread, Beer, Diaper\}(2)$ appear 2 times; whereas the following patterns $\{Beer, Butter\}(1)$, $\{Butter, Diaper\}(1)$, $\{Bread, Beer, Butter\}(1)$, $\{Bread, Butter, Diaper\}(1)$, $\{Beer, Butter, Diaper\}(1)$ and the whole set of items $\{Bread, Beer, Butter, Diaper\}(1)$ just appear one time.

1.2.5.2   Sequential Data

In many application fields gathered data is arranged in a sequential way and subsequent analysis to extract useful information for such data is performing by finding sequences of events that are present in at least a fraction of records. In this type of data, an event is defined as a collection of items that appear together and, therefore, they do not have a temporal ordering. In a formal way [3] and considering the set of n items in a dataset Ω in the form $I = \{i_1, i_2, ..., i_n\}$, an event e_j is defined as a non-empty unordered collection of items, i.e. $\{e_j = \{i_k, ..., i_m\} \subseteq I, 1 \le k, m \le n\}$. In this type of data, $T = \{t_1, t_2, ..., t_m\}$ is defined as a collection of transactions or sequence of events. Here, each transaction t_j is denoted as a sequence of events in the form $t_j = \langle e_1 \rightarrow ... \rightarrow e_n \rangle$ and each event e_i is described as an itemset $\{i_i, ..., i_j\}$ defined in the set of items $I \in \Omega$.

As a matter of example, let us consider the transaction or sequence of events $t_j = \langle \{Beer\}, \{Bread, Butter\}, \{Diaper\} \rangle$, which describes a customer who bought the item $Beer$, then $Bread$ and $Butter$ (not matter the temporal order) and, finally, $Diaper$. Taking this transaction as a baseline, a sequential pattern P_1 defined on this transaction is $P_1 = \langle \{Beer\}, \{Butter\} \rangle$ since any pattern P defined on a transaction is a subset of such a transaction keeping the order of the events. On the contrary, the pattern $P_2 = \langle \{Beer\}, \{Diaper, Butter\} \rangle$ is not a valid pattern on t_j since the second event ($\{Diaper, Butter\}$) is not a subset of any of the remain events of t_j.

1.2.5.3 Spatiotemporal Data

As previously defined, sequential data include features related to the time stamp, describing some sequential behaviour. The use of the time stamp can also be associated with different spacial features, giving rise to spatio-temporal data [3]. In spatiotemporal data, each data record is essentially a trajectory defined by the timestamp and the location so, in general, a dataset includes a set of trajectories of n moving objects. Each trajectory (data record) includes a sequence of points describing the coordinates (x_i, y_i) and a specific time stamp ts_i. Here, (x_i, y_i) is a location (longitude and latitude) and ts_i is the time when the location (x_i, y_i) was recorded.

This type of data has experienced a rapid development thanks to positioning technologies, sensor networks and social media. The analysis and extraction of useful information for such data may be essential in many application fields. For example, according to *Han et al.* [3], understanding animal movements may be of high-impact to study environmental challenges such as climate change, bio-diversity loss, and infectious diseases among others. The extraction of patterns on such data types is also of high interest to analyse traffic information and make the right decisions.

1.2.5.4 Graphs

Most of existing algorithms for data analysis are based on a flat and tabular representation where each data record includes a set of items. However, there are many application fields such as bioinformatics, cheminformatics or social network where data (3D protein structures, chemical compounds, XML files, etc) do not fit well in a flat representation. In this data representation, frequent graph patterns [3] are defined as subgraphs that frequently appear in a collection of graphs or in single large graph. Thus, frequent subgraphs are useful to characterize graphs, discriminate or cluster groups of graphs, etc. In this problem, a graph is formally defined as $G = (V, E, \mu, v)$ where V states for the finite set of nodes, $E \subseteq V \times V$ denotes the set of edges, $\mu : V \rightarrow L_V$ is a node labeling function, and $v : E \rightarrow L_E$ is an edge labeling function. Given two graphs $G_1 = (V_1, E_1, \mu_1, v_2)$ and $G_2 = (V_2, E_2, \mu_2, v_2)$, then G_1 is a subgraph of G_2 if $V_1 \subseteq V_2$, $E_1 \subseteq E_2$, $\forall u \in V_1 : \mu_1(u) = \mu_2(u)$, and $\forall (u, v) \in E_1 : v_1(u, v) = v_2(u, v)$. According to these definitions and given a labeled graph dataset $\Omega = \{G_1, G_2, ..., G_n\}$, a subgraph g is frequent if its support is no less than a minimum support threshold value. The support of a subgraph g is defined as $support(g) = |\Omega_g|/|\Omega|$, where $\Omega_g = \{G_i : g \subseteq G_i, G_i \in \Omega\}$.

1.2.5.5 High Utility Data

Pattern mining has attracted a lot of attention from researchers and users to find patterns that frequently appear in data or those that provide any unusual or interesting statistical behaviour. In some real-world applications such as the market basket analysis, such patterns are not always the most useful patterns since they may generate a low profit even though they are frequently purchased [48]. Unlike pattern mining on traditional datasets (tabular representation), high utility datasets include non-binary purchase quantities for the items in the transactions, and it also considers that all items are not always equally important.

A high utility dataset is represented as a finite set of distinct items $I = \{i_1, i_2, ..., i_n\}$ and each single item is associated with a positive integer that represents its utility or profit. A high utility dataset gathers a set of transactions and each transaction includes a subset of items from I. Additionally, each single item in each transaction has an internal utility that is obtained as the product of its utility and the purchase quantity of such item in the transaction. For the sake of clarification, let us define a set of profits for the items defined in a market basket dataset (Bread:1; Beer:5; Butter:2; and Diaper:5) and a sample dataset including the items and the number of purchases per item (see Table 1.3). Taking the first transaction, its utility is obtained as the product $1 \times 1 = 1$. As for the second transaction, its utility is obtained as the product $(1 \times 1) + (4 \times 5) = 21$. An itemset is therefore called high utility itemset if its utility is no less than a user-predefined minimum threshold.

1.2.5.6 Uncertain Data

Nowadays, the extraction of patterns of interest is commonly applied to real-life datasets including fields such as banking, bioinformatics, marketing, medical diagnosis, etc. In these real-life data, uncertainty may be present due to different reasons [3] including limitations in the understanding of reality and limitations of equipment that gathers data. When dealing with uncertain data, users may not be certain about the presence or absence of an item i_j in a transaction t_i. This uncertainty is usually expressed in terms of existential probability $P(i_j, t_i)$, which

Table 1.3 Sample market basket dataset including number of purchases per item

Items and purchases
{(Bread,1)}
{(Bread,1), (Beer,4)}
{(Bread,2), (Beer,1), (Diaper,6)}
{(Bread,1), (Butter,2)}
{(Bread,2), (Beer,4), (Butter,2), (Diaper,5)}
{(Beer,8)}
{(Beer,8), (Diaper,6)}
{(Bread,1), (Butter,1)}
{(Bread,1), (Beer,2), (Diaper,4)}
{(Bread,2)}

Table 1.4 Sample probabilistic market basket dataset of uncertain data

Items
{Bread:0.7}
{Bread:0.3, Beer:0.4}
{Bread:0.6, Beer:0.7, Diaper:0.9}
{Bread:0.5, Butter:0.5}
{Bread:0.3, Beer:0.4, Butter:0.2, Diaper:0.5}
{Beer:0.8}
{Beer:0.8, Diaper:0.9}
{Bread:0.1, Butter:0.6}
{Bread:0.5, Beer:0.5, Diaper:0.2}
{Bread:0.9}

indicates the likelihood of i_j being present in t_i, which ranges from a positive value close to 0 (i_j has an insignificantly low chance to be present) to a value of 1 (i_j is definitely present with no doubt). If these datasets are compared to traditional data (precise data), such precise datasets may be considered as uncertain data where the probability is 1 for all its items, that is, any item has a 100% likelihood of being present in a transaction.

The sample market basket dataset (see Table 1.1) defined in previous sections can be defined as an uncertain dataset including existential probabilities to each item (see Table 1.4). Here, the expected support of each item is computed as the sum of the expected probabilities of such item in each of the transactions in the database. In this sense, the expected support of each of the single items in data is computed as: $expected\_support(Bread) = 3.9$, $expected\_support(Beer) = 3.6$, $expected\_support(Butter) = 2.0$ and $expected\_support(Diaper) = 2.5$. When dealing with itemsets (those patterns comprising more than a single item), the probability of such itemset occurring in a specific transaction is computed as the product of the probabilities of each of its single items. Thus, the probability of the pattern $P = \{Bread, Beer\}$ occurring in the second transaction is equal to 0.12. The expected support of each itemset is therefore calculated in a similar way as previously described for single items. Thus, considering the aforementioned pattern

P, its expected support is calculated as $expected\_support(P) = 0 + 0.12 + 0.42 + 0 + 0.12 + 0 + 0 + 0 + 0.25 + 0 = 0.91$. Now, a pattern (an itemset) is frequent if its expected support is larger than a user-predefined threshold.

1.2.5.7 Multiple-Instance Data

Traditional datasets organize the information in a tabular manner were each row unequivocally identifies a single record including a set of elements or items that characterizes a specific object or ID. Sometimes, though, data information is ambiguous and a specific ID is described by an undefined number of different records, for example a market basket dataset where each customer is represented by a different number of records (one per purchase). This ambiguity cannot be handled by traditional pattern mining approaches [83], which provide a general description about homogeneity and regularity in the whole dataset. In fact, if the customers' ID is not considered during the mining process, it may happen that the extracted knowledge is biased against occasional customers.

In a formal way [57], and considering a database Ω comprising a set of n bags $\Omega = \{B^1, B^2, ..., B^n\}$, each bag B^j comprises an undetermined number of transactions, i.e. $B^j = \{t_1^j, ..., t_p^j\}$. Such bags represents different objects or IDs, for example each bag includes purchases of a specific customer. A pattern P occurs in a bag B^j if and only if P is a subset of at least one transaction t_i^j from B^j, i.e. $\exists t_i^j : t_i^j \in B^j \wedge P \subseteq t_i^j$. Thus, the support or frequency of a pattern in this data type is calculated as the number of bags in which P is included in at least one transaction, i.e. $support(P) = |\forall B^j \in \Omega, \exists t_i^j : t_i^j \in B^j \wedge P \subseteq t_i^j|$.

For the sake of improving the understanding, let us consider a dataset (see Table 1.5) comprising 4 different bags, each bag comprising a different number of records. Here, the pattern $P = \{Bread, Beer\}$ appear in 3 bags so its support is calculated as $support(P) = 3/4$, i.e. it is satisfied in 75% of the bags. In other words, the pattern P is satisfied in bags #1, #2, #3 even when in some bags the pattern P appears more than once (there are two records in the first bag including P).

Table 1.5 Sample market basket dataset organized into bags

Bags	Items
#1	{Bread}
	{Bread, Beer}
	{Bread, Beer, Diaper}
	{Bread, Butter}
#2	{Bread, Beer, Butter, Diaper}
	{Beer}
#3	{Beer, Diaper}
#4	{Bread, Butter}
	{Bread, Beer, Diaper}
	{Bread}

1.3 Supervised Descriptive Patterns

Data analysis is generally divided into two main groups, that is, predictive and descriptive tasks. Predictive tasks include techniques where models, typically induced from labeled data, are used to predict a target variable (labeled variable) of previously unseen examples. Descriptive tasks, however, aim at finding comprehensible patterns that denote any interesting behaviour on unlabeled data. Until recently, these techniques have been researched by two different communities: predictive tasks principally by the machine learning community [63], and descriptive tasks mainly by the data mining community [31]. However, different application domains sometimes require that both groups converge at some point giving rise to the concept of supervised descriptive pattern mining [66]. The main aim now is to understand an underlying phenomena (according to a target variable) and not to classify new examples.

Supervised descriptive discovery originally gathered three main tasks: contrast set mining [19], emerging pattern mining [28] and subgroup discovery [5]. Nowadays, however, there are many additional tasks that can be grouped under the supervised descriptive pattern mining concept since they provide any kind of descriptive knowledge from labeled data. In this section, the most important tasks in this regard are analysed.

1.3.1 Contrast Sets

According to *Dong et al.* [19], contrast patterns are itemsets that describe differences and similarities among groups the user wishes to contrast. Here, the groups under contrast are considered as subsets of a common dataset so each one may correspond to a target variable of the general dataset, or may represent subsets satisfying various conditions. It is important to remark the independence of such subsets, in such a way that a specific data record can only belong to a single subset. Thus, while classifiers place new data into a series of discrete categories, contrast set mining takes the category and provides a statistical evidence that identifies a record as a member of a class (target variable value).

In a formal way, *Bay and Pazzani* [8] defined a contrast set as a pattern P or conjunction of attributes and values that differ meaningfully in their distributions across groups. A special case of contrast set mining considers only two contrasting groups, looking for characteristics of one group discriminating it from the other and vice versa [66]. In general terms, contrast sets quantify the difference in support for each subset (group) S_i so P is therefore considered a contrast set if and only if $\exists ij :$ $max(|support(P, S_i) - support(P, S_j)|) \geq \alpha \in [0, 1]$, denoting $support(P, S_i)$ as the support of a pattern P on this subset. Since contrast sets were proposed to be able to compare what someone likes with something else, its representation is a keypoint, requiring to be interpretable, non redundant, potentially actionable and expressive.

One of the most well-known algorithms for this task is STUCCO (Search and Testing for Understandable Consistent Contrasts) [8], which discovers a set of contrast sets along with their supports on the groups. This algorithm includes a pruning mechanism to discard contrast sets if they fail a statistical test for independence with respect to the group variable. This concept is related to the one of redundant rules defined by *Webb* [86], which states that a rule $X \rightarrow Y$ is redundant and can be discarded if $\exists Z \in X : support(X \rightarrow Y) = support(X - Z \rightarrow Y)$. STUCCO discard rules $X \rightarrow Y$ if $\forall Z \subset X : confidence(Z \rightarrow Y) < confidence(X \rightarrow Y)$, the confidence being defined as $confidence(X \rightarrow Y) = support(X \rightarrow Y)/support(X)$, that is, the estimate of conditional probability $P(Y|X)$. As a result, a more specific contrast set must have higher confidence than any of its generalizations.

In 2003, *Webb et al.* [87] defined the mining of contrast sets as a special case of the more general rule learning task. Here, they defined the contrast set as the antecedent of a rule $X \rightarrow Y$, whereas the consequent includes those characteristics of the S_i group (subset) for which it is in contrast with the S_j group. In this sense, a standard association rule mining algorithm [93] in which the consequent is restricted to the variable (or variables) that denotes the membership to the groups can be used for mining contrast sets. In 2005, *Hilderman and Peckham* [35] proposed an approach called CIGAR (ContrastIng Grouped Association Rules), which is based on a set of statistical tests for redundant rules and it also considers a minimum support value.

Contrast set mining has been applied to different fields, including retail sales data [87]; design of customized insurance programs [89]; identification of patterns in synchrotron x-ray data that distinguish tissue samples of different forms of cancerous tumor [75]; classify between two groups of brain ischaemia patients [41].

1.3.2 Emerging Patterns

Emerging patterns were defined by *Dong et al.* [20] as a data mining task that searches discriminative patterns whose support increases significantly from one dataset to another. This task is close to the one of mining contrast sets so emerging patterns can be viewed as contrast patterns between two kind of data whose support changes significantly between the two data types. These patterns use an understandable representation to describe discriminative behavior between classes or emerging trends amongst datasets.

In a formal way, given a pattern P defined on two datasets Ω_1 and Ω_2, this pattern P is denoted as an emerging pattern if its support on Ω_1 is significantly higher than its support on Ω_2 or vice versa. This difference in support is quantified by the growth rate, that is, the ratio of the two supports ($support(P, \Omega_1)/support(P, \Omega_2)$ or $support(P, \Omega_2)/support(P, \Omega_1)$), and values greater than 1 denotes an emerging pattern. Here, the growth rate is 1 if $support(P, \Omega_1) = support(P, \Omega_2)$ and 0 if both P does not satisfy any record from either Ω_1 or Ω_2. In those cases where

$support(P, \Omega_1) = 0$ and $support(P, \Omega_2) \neq 0$, or $support(P, \Omega_1) \neq 0$ and $support(P, \Omega_2) = 0$, then the growth rate is defined as ∞. Some authors [66] denoted emerging patterns as association rules with an itemset (the pattern P) in the antecedent and a fixed consequent (the dataset), that is, $P \rightarrow \Omega$. Hence, the growth rate is calculated as $confidence(P \rightarrow \Omega_1)/confidence(P \rightarrow \Omega_2) = confidence(P \rightarrow \Omega_1)/(1 - confidence(P \rightarrow \Omega_1))$.

The main problem of mining emerging patterns is that it is a non-convex problem [84] so more general patterns may have lower growth rate values than more specific patterns. For example, a pattern with a 0.1 relative support value in one data set and 0.01 in the other is better than a pattern with relative support value of 0.7 in one data set and 0.1 in the other $(0.1/0.01 > 0.7/0.1)$. As a result, a reduction in the search space by using minimum support values, as traditional pattern mining algorithms do, is not a recommendation. Many different researchers have focused their studies on reducing the large number of solutions. In 2003, *Fan et al.* [23] proposed an algorithm, called iEPMiner, to select interesting emerging patterns defined as those which have a minimum support; a minimum growth rate and larger than its subset; and which are highly correlated according to common statistical measures. *Soulet et al.* [78] proposed condensed representations of emerging patterns similar to the condensed representations of frequent closed patterns. The proposed approach enabled the implementation of powerful pruning criteria during the mining process.

Emerging patterns have been mainly applied to the field of bioinformatics and, more specifically, to microarray data analysis [46]. *Li et al.* [44] focused their studies on mining emerging patterns to analyse genes related to colon cancer. Finally, emerging patterns have been also applied to analyse customer behaviour [76] to discover unexpected changes in the customers' habits.

1.3.3 Subgroup Discovery

Subgroup discovery [5] is a really interesting technique for mining supervised descriptive patterns, which aims at identifying a set of patterns of interest according to their distributional unusualness with respect to a certain property of interest (target concept). The concept of subgroup discovery was first introduced by *Klösgen* [38] and *Wrobel* [90] as follows:

"Given a population of individuals (customers, objects, etc.) and a property of those individuals that we are interested in, the task of subgroup discovery is to find population subgroups that are statistically most interesting for the user, e.g. subgroups that are as large as possible and have the most unusual statistical characteristics with respect to a target attribute of interest."

Subgroup discovery combines features of descriptive and predictive tasks [11] and uncovers explicit subgroups via single and simple rules, that is, having a clear structure and few variables [34]. These rules, in the form $P \rightarrow Target$, denote an unusual statistical distribution of P (pattern including a set of features) with respect to the target concept or variable of interest. In this supervised descriptive task, target concepts have been studied by considering either binary and nominal domains, that is, properties of interest with a finite number of possible values. Nevertheless, target concepts on continuous domains have received increasing attention in recent years [5].

The interest of the rules that form subgroups are quantified according to a wide variety of metrics [21]. These quality measures can be divided into two main groups, depending on the target variable. As for discrete target variables, a recent taxonomy of metrics was proposed, including four main types: measures of complexity (interpretability of the discovered subgroups and simplicity of the extracted knowledge); generality (quality of the subgroups according their coverage); precision (reliability of each subgroup); and interest (significance and unusualness of the subgroups). There is also an additional group of measures that are hybrid metrics defined to obtain a tradeoff between generality, interest and precision. Regarding numerical target variables, quality measures are categorized with respect to the used data characteristics as follows [43]: mean-based interestingness measures, variance-based measures, median-based measures and distribution-based measures.

First algorithms developed for the subgroup discovery task were defined under the names EXPLORA [38] and MIDOS [90]. These algorithms were described as extensions of classification algorithms and they used decision trees to represent the knowledge. CN2-SD [40] is another example of a subgroup discovery approach based on classification algorithms. On the contrary, many other authors [83] considered the subgroup discovery problem as an extension of association rule mining, including Apriori-SD [36] and SD-Map [6] among others. Recently, though, the subgroup discovery problem is being solved by means of evolutionary algorithms to reduce the computational time and to increase the interpretability of the resulting model, including proposals such as CGBA-SD [52] and MESDIF [14], among others.

Finally, it is important to highlight that subgroup discovery has been used in numerous real-life applications such as the analysis of coronary heart disease [27]; census data [39]; marketing [17]; students' attitudes in high school [65]; etc.

1.3.4 Class Association Rules

The concept of association rule was proposed by *Agrawal et al.* [4] as a way of describing correlations among items within a pattern. Let P be a pattern defined in a database Ω as a subset of items $I = \{i_1, ..., i_n\} \in \Omega$, i.e. $P \subseteq I$, and let also consider X and Y subsets of P, i.e. $X \subset P \subseteq I$ and $Y = P \setminus X$. An association rule is formally defined as an implication of the form $X \rightarrow Y$ where X and Y do not have any common item, i.e. $X \cap Y = \emptyset$. The meaning of an association rule is that if the antecedent X is satisfied, then it is highly probable that the consequent Y is also satisfied, i.e. one of the sets leads to the presence of the other set [93].

The resulting set of association rules that is produced when applying any association rule mining algorithm can be decomposed into a more specific set including rules with a specific attribute in the consequent. These rules, known as class association rules (CARs) are formally defined as implications of the form $X \rightarrow y$, X being a subset of items in data $X \subseteq I$ and y defined as the target variable or class attribute. This subset of association rules may include a total of $(2^k - 1) \times t$ solutions for a dataset containing k items and a target variable y comprising t different values. Thus, the resulting set of solutions is much smaller than the one of general association rules, which comprises up to $3^k - 2^{k+1} + 1$ solutions for a dataset comprising k items.

Generally speaking, CARs are target-constraint rules in the form $P \rightarrow Target$, so this task is highly related to the one of subgroup discovery. The main difference lies in the fact that subgroup discovery searches for a reduced set of reliable rules that is easily interpretable, simple, and covers as many examples as possible. On the contrary, the task of mining CARs is more focused on the extraction of any rule that satisfies some minimum threshold values (support and confidence are generally used as quality measures). Finally, it is important to highlight that, since its definition [49], CARs have gained an increasing attention due to their understandability and many research studies have considered these rules not only for descriptive purposes [54] but also to form accurate classifiers [49].

Focusing on descriptive purposes, the aim of mining CARs is to find a set of items that properly describe the target variable or variable of interest. A major feature of CARs is that the simple fact of considering a single item or target variable in the consequent may help in improving the understanding of the provided knowledge [83]. In fact, this type of association rules are more frequently used than those including a large number of items in the consequent (general association rules). Some of the most well-known algorithms in association rule mining (Apriori [4] and FP-Growth [32]) can be easily modified to be used in the mining of CARs. Additionally, a wide number of approaches have been specifically proposed for this task [51, 53, 59, 62].

As for predictive tasks, *Liu et al.* [49] used CARs to form accurate and interpretable classifiers. Here, thanks to CARs and by considering the class as the consequent of the rule, it is possible to take advantage of reliable and interesting association rules to form accurate and very interpretable classifiers. In order to obtain this kind of classifiers, the complete set of CARs should be first discovered from the training data set and a subset of such rules is then selected to form the classifier. The selection of such a subset can be accomplished in many ways, and first approaches (CBA [49] or CMAR [45]) were based on selecting rules by using a database coverage heuristic [49]. Nowadays, a wide number of approaches can be found in literature [7, 80, 81, 91] for mining CARs, which have been correctly applied to numerous problems, including traffic load prediction [94]; analysis of students' performance on learning management systems [54]; gene expression classification [37]; among others.

1.3.5 Exceptional Models

Exceptional model mining [42] is defined as a multi-target generalization of subgroup discovery [34], which is the most extensively studied form of supervised descriptive rule discovery [66]. This task searches for interesting data subsets on predefined target variables and describes reasons to understand the cause of the unusual interactions among the targets.

Leman et al. [42] formally described the exceptional model mining task as the discovery of subgroups where a model (based on the target variables) fitted to the subgroup is substantially different from the same model fitted to the entire database. In other words, given a dataset Ω including a collection of records $r \in \Omega$ in the form $r = \{d_1, ..., d_k, t_1, ..., t_m\} : k, m \in N^+$, $D = \{d_1, ..., d_k\}$ being the set of descriptors, and $T = \{t_1, ..., t_m\}$ stating for the target variables, then the aim of the exceptional model mining task is to extract a subset of data $G_D = \{\forall r^i \in \Omega :$ $D(d_1^i, ..., d_k^i) = 1\}$, corresponding to a description given by a set of descriptive attributes $D(d_1, ..., d_k)$, and denoting an unusual interaction on two specific target variables t_x and t_y from T. This interaction was originally quantified in terms of the *Pearson's* standard correlation coefficient ρ between t_x and t_y for both the subset ρ^{G_D} and the whole dataset ρ^Ω (or the complement $G_D^C \equiv \Omega \setminus G_D$, denoted as $\rho^{G_D^C}$). Figure 1.6 illustrates a sample exceptional model defined on two target variables (X and Y). In this example, P states for the pattern that includes the set of descriptors that characterize the subset G_P, whereas $G_P^C \equiv \Omega \setminus G_P$. Here, both G_P and G_P^C denote an unusual interaction on X and Y.

According to *Duivesteijn et al.* [22], there are two main schools of thought in the exceptional model mining community on how to overcome this task. The first one restricts descriptive attributes in the dataset to be nominal and imposes the Apriori [4] constraint so the search space can be explored exhaustively by considering traditional subgroup discovery methods [34]. The second is related to the use of an heuristic search, enabling descriptive attributes to be numeric as well. Here, a beam search strategy is taken [22], which performs a level-wise search— the best descriptions according to a quality measure are selected on each level. Recently, the exceptional model mining task was defined as a halfway between association rule mining and exceptional model mining, giving rise to the mining of exceptional relationships [55]. This subtask enables the discovery of accurate relations among patterns [83] where the subgroup fitted to a specific set of target features is significantly different to its complement.

Exceptional model mining can be considered in many application fields as it was demonstrated by *Duivesteijn et al.* [22], where the problem was applied to the analysis of the housing price per square meter, mammals that appear only in a very small subarea of Europe, or how the expression of a gene may be influenced by the expression of other genes. Exceptional association rules, on the contrary, has been applied to the analysing unusual and unexpected relationships between quality measures in association rule mining [55].

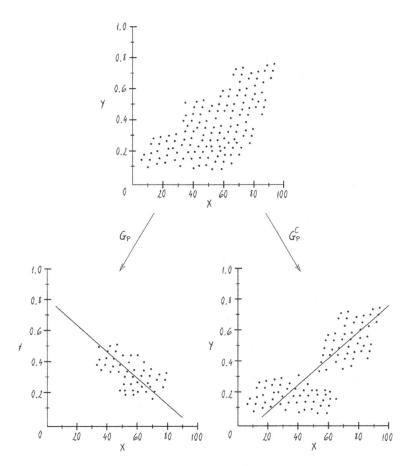

Fig. 1.6 Unusual interaction on two target variables (X and Y) described by the pattern P (set of descriptors)

1.3.6 Other Forms of Supervised Descriptive Patterns

Supervised descriptive pattern mining includes tasks aiming at providing any kind of descriptive knowledge from labeled data [66]. Techniques proposed for these tasks usually seek descriptors or concepts that characterize and/or discriminate a target variable (or multiple targets). Sometimes, these descriptors can be compacted into a reduced set that formally reveals the same information as the whole set, as it is the case of closed patterns [3]. In this regard, similarly to unlabeled data, closed patterns can be used to compress information from labeled data, which is really useful for classification and discrimination purposes [29].

An important form of looking for supervised descriptive patterns is bump hunting [24], which looks for subregions of the space of variables within which the value of the target variable is considerably larger (or smaller) than its average value over the whole search space. Bump hunting is therefore a specific case of subgroup discovery where the interest is not focused on metrics of complexity, generality, precision, and interest. Instead, this task aims at mining subsets (subgroups) that highly characterize a target variable (defined in a discrete domain).

Sometimes, though, the target variable is not defined as a discrete variable but as an undiscretized quantitative variable. This is the case of impact rules [85], which are defined as a specific type of class association rules where the consequent is a target variable defined in a continuous domain. Unlike class association rules, impact rules express the consequent together with different measures including coverage, mean, variance, maximum value, minimum value, among others. Thus, an impact rule is represented as a class association rule in the form $X \rightarrow Target\{coverage = V_1, mean = V_2, variance = V_3, maximum\_value = V_4, ...\}$ where V_1 is the percentage of records in which the set X is true; V_2 is the mean of values of the $Target$ variable for which X is true; V_3 is the variance of the $Target$ values for which X is true; etc.

Another important and different form of supervised descriptive pattern mining is known as change mining. It aims at modelling, monitoring, and interpreting changes in the patterns that describe an evolving domain over many temporally ordered data sets [10]. The final objective of this task is to produce knowledge that enables any anticipation of the future so it is particularly helpful in domains in which it is crucial to detect emerging trends as early as possible and which require to make decisions in anticipation rather than in reaction.

When considering data mining techniques for dealing with historical decision records, there is a lack of care about minorities or protected-by-law groups. On this matter, *Turini et al.* [73] proposed the notion of discrimination discovery as a criterion to identify and analyse the potential risk of discrimination. In this problem, given a dataset of historical decision records, a set of potentially discriminated groups, and a criterion of unlawful discrimination, the aim is to find subsets of the decision records, called context, and potentially discriminated groups within each context for which the criterion of unlawful discrimination hold. The discrimination discovery task is highly related to the one of extracting contextual relationships among patterns (a form of class association rules). Here, it is important to describe that real-world data usually comprise features whose interpretation depends on some contextual information [59] so it is crucial to analyse such features to obtain the right meaning. As a matter of example, the concept of strength is completely different for a child, an elderly person or even a teenager. However, a person is assigned to any of these contexts at different times of his/her life.

1.4 Scalability Issues

Nowadays, the amount of data that races through the Internet every second is even higher than the total stored two decades ago. In this sense, the increasing interest in data storage has given rise to increasingly large datasets [61], which may exceed the capacity of traditional techniques with respect to memory consumption and performance in a reasonable quantum of time. When focusing on the pattern mining task, the problem turns extremely complex since the total number of patterns that might be generated from k single items is equal to $2^k - 1$, that is, it exponentially increases with the number of items. In this regard, it is primordial to carry out the mining process by pruning the search space [3] or considering new heuristic approaches [83] that do not require an exploration of the whole search space. Additionally, many research studies have been focused on minimizing both the runtime and memory requirements through the application of new data structures [56] as well as emerging parallel techniques such as multi-core processors, graphic processing units (GPUs) [13] and the MapReduce framework [16].

1.4.1   Heuristic Approaches

The extraction of patterns of interest can be formulated as a combinatorial optimization problem aiming at obtaining a set of optimal patterns according to some specific quality measures [93]. The use of an evolutionary methodology has been studied in the pattern mining field by a large number of researchers [83] since it enables the computational and memory requirements to be tackled. In any evolutionary algorithm, the final performance is highly related to the genetic operators used during the evolutionary process, and these operators enable to guide the searching process to promising regions of the search space and to discard those areas that seem to be meaningless due to solutions around them are far to be optimal.

Despite the fact that the mining of patterns of interest has been studied by means of different metaheuristics [18, 83], most of the existing algorithms in this field were developed through genetic algorithms (GAs). In any GA, it is obtained that its efficiency as well as the quality of its solutions are highly dependent on the population size and the number of generations the algorithm takes to converge (to reach the optimum). It should be noted that both a large population size and a high number of generations imply an increase in the probability of analysing a higher portion of the search space. Finally, it is important to highlight that some recent studies on the pattern mining field have included the use of context-free grammars into the mining process so they enable subjective knowledge to be introduced into the mining process as well as a reduction in the space of solutions (only those solutions satisfying the grammars are feasible) [59].

1.4.2   New Data Structures

First proposals for speeding up the process of mining patterns were based on efficient representations of the lattice of frequent itemsets [3]. The most well-known example of this is the FP-Growth algorithm [32] which represents the whole dataset into a prefix-tree structure (it was previously described in Sect. 1.2.1). In this tree structure, each node represents a single item, whereas the path represents the set of items that form the pattern. FP-Growth is a highly efficient algorithm since it requires to analyse the dataset only once and no further passes over the dataset are required. Nevertheless, this algorithm still requires large amounts of memory and a high computational time for analysing all the feasible patterns in huge datasets. In this regard, many researchers are working nowadays on the proposal of new data structures that simplify and reorganize data items in order to reduce the size and the time required to access to the information.

All the proposed data structures are based on the fact that records in data tend to share items so these records may be sorted and compressed. Some authors have employed techniques such as inverted index mapping [50] or run length encoding compression [1] to re-organize items and records in data so the whole dataset can fit into main memory. It is therefore possible to reduce the memory storage requirements by sorting data in a special way by using a shuffling strategy like, for example, the hamming distance. *Luna et al.* [56] proposed the use of the hamming distance as a similarity metric so data records are grouped in order to minimize the number of changes in the values of the items. In other words, the aim was to cluster those records with similar items so they could be compressed by means of a inverted index mapping procedure, enabling lower memory requirements and speeding up the data accesses.

1.4.3   Parallel Computing

Many different methodologies have been developed for solving the arduous problem of mining patterns of interest including efficient exhaustive search approaches [3], new data representations [32, 56], and different evolutionary proposals [83]. However, when dealing with extremely large datasets, the complexity in terms of runtime and memory requirement turns unmanageable. This gave rise to the idea of using parallel approaches [30] by considering either multi-core processors and graphic processing units (GPUs).

In 2009, *Fang et al.* [88] proposed the two first parallel approaches for mining frequent patterns, which were based Apriori, the most well-known exhaustive search algorithm in this field. Then, *Adil et al.* [2] proposed the use of a GPU architecture for mining association rules. In this proposal, the mining process was designed into two different steps: (*a*) frequent itemsets were computed through different thread blocks; (*b*) the set of frequent itemsets previously obtained were sent back to the CPU to generate associations among such itemsets. According to the authors, the major drawback of this algorithm was the cost of the CPU/GPU communications.

Cui et al. [15] proposed a frequent pattern mining algorithm based on CUDA (Computer Unified Device Architecture), a parallel computing architecture developed by NVIDIA that allows programmers to take advantage of the computing capacity of NVIDIA GPUs in a general purpose manner. The use of GPUs and the CUDA programming model enables the solutions to be evaluated in a massively parallel way, thus reducing the computational time required by traditional pattern mining approaches. Due to high scalable approaches that can be designed with CUDA, some authors have considered not only exhaustive search approaches but also proposals based on different metaheuristics [13]. According to the authors, one of the most time-consuming process in any evolutionary algorithm is the evaluation phase where the higher the number of records in data to be checked, the higher the computational required by this phase. However, this process tends to be independent from solution to solution so it provides a high degree of parallelism since each solution may be computed in a concurrent way.

1.4.4   MapReduce Framework

MapReduce [16] is one of the most important and well-known emerging technologies for parallel processing and intensive computing. This programming model offers a simple and robust way of writing distributed programs, specially those designed for tackling large datasets that require data-intensive computing. MapReduce programs include two main phases defined by the programmer: Map and Reduce. During the first phase, the input data is divided into subsets that are processed in a parallel way, producing a set of $\langle k, v \rangle$ pairs. Then, the Reduce phase is responsible for combining all the keys k to produce final values.

For a better understanding, let us consider a simple example about a MapReduce process for mining patterns (Table 1.1 is taken as baseline). According to the MapReduce framework (see Fig. 1.7), the original dataset is split into subsets of

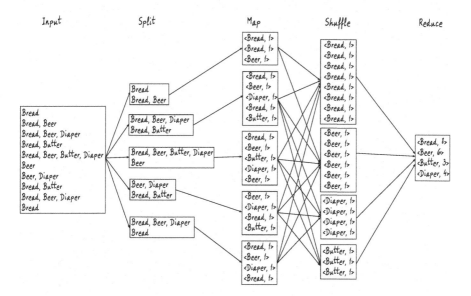

Fig. 1.7 MapReduce framework applied to the sample dataset shown in Table 1.1

data and each mapper receives one of these subsets. Here, each mapper iterates over each item, producing a $\langle k, v \rangle$ pair for each feasible item that can be obtained in each subset. k states for the name of the item in the dataset, and v means the number of times that k appear in the subset. Then, a shuffling procedure is carried out before the reduce phase. This procedure is responsible for ordering all the $\langle k, v \rangle$ pairs obtained by the mappers so all those $\langle k, v \rangle$ pairs that share the same key k are grouped. Finally, the reduce phase calculates the sum of v for all the pairs grouped under the same k. As a result, the MapReduce process obtains the support values for each single item in data.

Similarly to existing sequential proposals for mining patterns [3, 83], which were based either on exhaustive search models or evolutionary algorithms, MapReduce has been properly used for mining patterns of interest on such methodologies. The main difference lies on the use of a MapReduce framework that parallelizes the mining process in a similar way as previously illustrated (see Fig. 1.7). Focusing on exhaustive search methodologies, *Moens et al.* [64] proposed the Dist-Eclat algorithm as well as the BigFIM approach. Recently, *Luna et al.* [58] proposed a series of algorithms that can be divided into three main groups (no pruning strategy, pruning strategy, and condensed representations of frequent patterns). In the first group, two algorithms were proposed (AprioriMR and Iterative AprioriMR). The second group includes two algorithms (AprioriMR and Top AprioriMR) that prune the search space by means of the well-known anti-monotone property. The third group includes an algorithm for mining condensed representations of frequent patterns is proposed (Maximal AprioriMR). It is noteworthy to mention that techniques for mining supervised local patterns have also been considered through exhaustive search methodologies on MapReduce, including subgroup discovery [69] and class association rules [68]. Finally, focusing on methodologies based on evolutionary computation, some authors have recently dealt with the supervised descriptive pattern mining problem by considering the MapReduce paradigm [70].

References

1. D.J. Abadi, S. Madden, M. Ferreira, Integrating compression and execution in column-oriented database systems, in *Proceedings of the ACM SIGMOD International Conference on Management of Data, SIGMOD Conference*, Chicago, Illinois (2006), pp. 671–682
2. S.H. Adil, S. Qamar, Implementation of association rule mining using CUDA, in *Proceedings of the 2009 International Conference on Emerging Technologies, ICET 2009*, Islamabad (2009), pp. 332–336
3. C.C. Aggarwal, J. Han, *Frequent Pattern Mining* (Springer International Publishing, Cham, 2014)
4. R. Agrawal, T. Imielinski, A.N. Swami, Mining association rules between sets of items in large databases, in *Proceedings of the 1993 ACM SIGMOD International Conference on Management of Data, SIGMOD Conference '93*, Washington, DC (1993), pp. 207–216
5. M. Atzmueller, Subgroup discovery - advanced review. WIREs Data Min. Knowl. Disc. **5**, 35–49 (2015)

6. M. Atzmueller, F. Puppe, SD-Map – a fast algorithm for exhaustive subgroup discovery, in *Proceedings of the 10th European Symposium on Principles of Data Mining and Knowledge Discovery, PKDD '06*, Berlin (2006), pp. 6–17

7. E. Baralis, S. Chiusano, P. Garza, A lazy approach to associative classification. IEEE Trans. Knowl. Data Eng. **20**(2), 156–171 (2008)

8. S.D. Bay, M.J. Pazzani, Detecting group differences: mining contrast sets. Data Min. Knowl. Disc. **5**(3), 213–246 (2001)

9. M.J. Berry, G. Linoff, *Data Mining Techniques: For Marketing, Sales, and Customer Support* (Wiley, New York, 2011)

10. M. Boettcher, Contrast and change mining. WIREs Data Min. Knowl. Discovery **1**(3), 215–230 (2011)

11. O. Bousquet, U. Luxburg, G. Ratsch, *Advanced Lectures On Machine Learning* (Springer, Berlin, 2004)

12. S. Brin, R. Motwani, J.D. Ullman, S. Tsur, Dynamic itemset counting and implication rules for market basket data, in *Proceedings of the 1997 ACM SIGMOD International Conference on Management of Data, SIGMOD '97*, Tucson, Arizona (ACM, New York, 1997), pp. 255–264

13. A. Cano, J. M. Luna, S. Ventura, High performance evaluation of evolutionary-mined association rules on gpus. J. Supercomput. **66**(3), 1438–1461 (2013)

14. C.J. Carmona, P. González, M.J. del Jesus, M. Navío-Acosta, L. Jiménez-Trevino, Evolutionary fuzzy rule extraction for subgroup discovery in a psychiatric emergency department. Soft Comput. **15**(12), 2435–2448 (2011)

15. Q. Cui, X. Guo, Research on parallel association rules mining on GPU, in *Proceedings of the 2nd International Conference on Green Communications and Networks, GCN 2012*, Gandia (2012), pp. 215–222

16. J. Dean, S. Ghemawat, MapReduce: simplified data processing on large clusters. Commun. ACM **51**(1), 107–113 (2008)

17. M.J. del Jesus, P. Gonzalez, F. Herrera, M. Mesonero, Evolutionary fuzzy rule induction process for subgroup discovery: a case study in marketing. IEEE Trans. Fuzzy Syst. **15**(4), 578–592 (2007)

18. M.J. del Jesús, J.A. Gámez, P. González, J.M. Puerta, On the discovery of association rules by means of evolutionary algorithms. Wiley Interdiscip. Rev.: Data Min. Knowl. Disc. **1**(5), 397–415 (2011)

19. G. Dong, J. Bailey (eds.), *Contrast Data Mining: Concepts, Algorithms, and Applications* (CRC Press, Boca Raton, 2013)

20. G. Dong, J. Li, Efficient mining of emerging patterns: discovering trends and differences, in *Proceedings of the 5th ACM SIGKDD International Conference on Knowledge Discovery and Data Mining (KDD-99)*, New York (1999), pp. 43–52

21. W. Duivesteijn, A.J. Knobbe, Exploiting false discoveries - statistical validation of patterns and quality measures in subgroup discovery, in *Proceedings of the 11th IEEE International Conference on Data Mining, ICDM 2011*, Vacouver, BC (2011), pp. 151–160

22. W. Duivesteijn, A. Feelders, A.J. Knobbe, Exceptional model mining - supervised descriptive local pattern mining with complex target concepts. Data Min. Knowl. Disc. **30**(1), 47–98 (2016)

23. H. Fan, K. Ramamohanarao, A bayesian approach to use emerging patterns for classification, in *Proceedings of the 14th Australasian Database Conference, ADC '03*, Adelaide (2003), pp. 39–48

24. J.H. Friedman, N.I. Fisher, Bump hunting in high-dimensional data. Stat. Comput. **9**(2), 123–143 (1999)

25. J. Fürnkranz, From local to global patterns: evaluation issues in rule learning algorithms, in *International Seminar on Local Pattern Detection*, Dagstuhl Castle (Springer, Berlin, 2004), pp. 20–38

26. J. Gama, *Knowledge Discovery from Data Streams*. Chapman and Hall/CRC Data Mining and Knowledge Discovery Series (CRC Press, Boca Rotan, 2010)

27. D. Gamberger, N. Lavrac, Expert-guided subgroup discovery: methodology and application. J. Artif. Intell. Res. **17**(1), 501–527 (2002)
28. A.M. García-Vico, C.J. Carmona, D. Martín, M. García-Borroto, M.J. del Jesus, An overview of emerging pattern mining in supervised descriptive rule discovery: taxonomy, empirical study, trends and prospects. Wiley Interdiscip. Rev. Data Min. Knowl. Disc. **8**(1) (2018)
29. G.C. Garriga, P. Kralj, N. Lavrač, Closed sets for labeled data. J. Mach. Learn. Res. **9**, 559–580 (2008)
30. T. George, M. Nathan, M. Wagner, F. Renato, Tree projection-based frequent itemset mining on multi-core CPUs and GPUs, in *Proceedings of the 22nd International Symposium on Computer Architecture and High Performance Computing, SBAC-PAD 2010*, Petrópolis (2010), pp. 47–54
31. J. Han, M. Kamber, *Data Mining: Concepts and Techniques* (Morgan Kaufmann, Burlington, 2000)
32. J. Han, J. Pei, Y. Yin, R. Mao, Mining frequent patterns without candidate generation: a frequent-pattern tree approach. Data Min. Knowl. Disc. **8**, 53–87 (2004)
33. J. Han, H. Cheng, D. Xin, X. Yan, Frequent pattern mining: current status and future directions. Data Min. Knowl. Disc. **15**(1), 55–86 (2007)
34. F. Herrera, C.J. Carmona, P. González, M.J. del Jesus, An overview on subgroup discovery: foundations and applications. Knowl. Inf. Syst. **29**(3), 495–525 (2011)
35. R.J. Hilderman, T. Peckham, A statistically sound alternative approach to mining contrast sets, in *Proceedings of the 4th Australasian Data Mining Conference (AusDM)*, Sydney (2005), pp. 157–172
36. B. Kavšek, N. Lavrač, APRIORI-SD: adapting association rule learning to subgroup discovery. Appl. Artif. Intell. **20**(7), 543–583 (2006)
37. K. Kianmehr, M. Kaya, A.M. ElSheikh, J. Jida, R. Alhajj, Fuzzy association rule mining framework and its application to effective fuzzy associative classification. Wiley Interdiscip. Rev.: Data Min. Knowl. Disc. **1**(6), 477–495 (2011)
38. W. Klösgen, Explora: a multipattern and multistrategy discovery assistant. in *Advances in Knowledge Discovery and Data Mining*, ed. by U.M. Fayyad, G. Piatetsky-Shapiro, P. Smyth, R. Uthurusamy (American Association for Artificial Intelligence, Menlo Park, 1996), pp. 249–271
39. W. Klosgen, M. May, J. Petch, Mining census data for spatial effects on mortality. Intell. Data Anal. **7**(6), 521–540 (2003)
40. N. Lavrač, B. Kavšek, P. Flach, L. Todorovski, Subgroup discovery with cn2-sd. J. Mach. Learn. Res. **5**, 153–188 (2004)
41. N. Lavrac, P. Kralj, D. Gamberger, A. Krstacic, Supporting factors to improve the explanatory potential of contrast set mining: analyzing brain ischaemia data, in *Proceedings of the 11th Mediterranean Conference on Medical and Biological Engineering and Computing (MEDICON-07)*, Ljubljana (2007), pp. 157–161
42. D. Leman, A. Feelders, A.J. Knobbe, Exceptional model mining, in *Proceedings of the European Conference in Machine Learning and Knowledge Discovery in Databases, ECML/PKDD 2008*, Antwerp, vol. 5212 (Springer, Berlin, 2008), pp. 1–16
43. F. Lemmerich, M. Atzmueller, F. Puppe, Fast exhaustive subgroup discovery with numerical target concepts. Data Min. Knowl. Disc. **30**(3), 711–762 (2016)
44. J. Li, L. Wong, Identifying good diagnostic gene groups from gene expression profiles using the concept of emerging patterns. Bioinformatics **18**(10), 1406–1407 (2002)
45. W. Li, J. Han, J. Pei, CMAR: accurate and efficient classification based on multiple class-association rules, in *Proceedings of the 1st IEEE International Conference on Data Mining, ICDM 2001*, San Jose, CA (2001), pp. 369–376
46. J. Li, H. Liu, J.R. Downing, A.E. Yeoh, L. Wong, Simple rules underlying gene expression profiles of more than six subtypes of acute lymphoblastic leukemia (ALL) patients. Bioinformatics **19**(1), 71–78 (2003)

47. Y. Li, A. Algarni, N. Zhong, Mining positive and negative patterns for relevance feature discovery, in *Proceedings of the 16th ACM SIGKDD International Conference on Knowledge Discovery and Data Mining, KDD '10*, Washington, DC (ACM, New York, 2010), pp. 753–762

48. J.C.-W. Lin, W. Gan, P. Fournier-Viger, T.-P. Hong, V.S. Tseng, Efficient algorithms for mining high-utility itemsets in uncertain databases. Knowl. Based Syst. **96**, 171–187 (2016)

49. B. Liu, W. Hsu, Y. Ma, Integrating classification and association rule mining, in *Proceedings of the Fourth International Conference on Knowledge Discovery and Data Mining, KDD-98*, New York City, New York (1998), pp. 80–86

50. R.W.P. Luk, W. Lam, Efficient in-memory extensible inverted file. Inf. Syst. **32**(5), 733–754 (2007)

51. J.M. Luna, J.R. Romero, S. Ventura, G3PARM: a grammar guided genetic programming algorithm for mining association rules, in *Proceedings of the IEEE Congress on Evolutionary Computation, IEEE CEC 2010*, Barcelona (2010), pp. 2586–2593

52. J.M. Luna, J.R. Romero, C. Romero, S. Ventura, On the use of genetic programming for mining comprehensible rules in subgroup discovery. IEEE Trans. Cybern. **44**(12), 2329–2341 (2014)

53. J.M. Luna, J.R. Romero, S. Ventura, On the adaptability of G3PARM to the extraction of rare association rules. Knowl. Inf. Syst. **38**(2), 391–418 (2014)

54. J.M. Luna, C. Romero, J.R. Romero, S. Ventura, An evolutionary algorithm for the discovery of rare class association rules in learning management systems. Appl. Intell. **42**(3), 501–513 (2015)

55. J.M. Luna, M. Pechenizkiy, S. Ventura, Mining exceptional relationships with grammar-guided genetic programming. Knowl. Inf. Syst. **47**(3), 571–594 (2016)

56. J.M. Luna, A. Cano, M. Pechenizkiy, S. Ventura, Speeding-up association rule mining with inverted index compression. IEEE Trans. Cybern. **46**(12), 3059–3072 (2016)

57. J.M. Luna, A. Cano, V. Sakalauskas, S. Ventura, Discovering useful patterns from multiple instance data. Inf. Sci. **357**, 23–38 (2016)

58. J.M. Luna, F. Padillo, M. Pechenizkiy, S. Ventura, Apriori versions based on mapreduce for mining frequent patterns on big data. IEEE Trans. Cybern. 1–15 (2018). Online first. https://doi.org/10.1109/TCYB.2017.2751081

59. J.M. Luna, M. Pechenizkiy, M.J. del Jesus, S. Ventura, Mining context-aware association rules using grammar-based genetic programming. IEEE Trans. Cybern. 1–15 (2018). Online first. https://doi.org/10.1109/TCYB.2017.2750919

60. M. Martinez-Ballesteros, I.A. Nepomuceno-Chamorro, J.C. Riquelme, Inferring gene-gene associations from quantitative association rules, in *Proceedings of the 11th International Conference on Intelligent Systems Designe and Applications, ISDA 2011*, Cordoba (2011), pp. 1241–1246

61. V. Marx, The big challenges of big data. Nature **498**(7453), 255–260 (2013)

62. J. Mata, J.L. Alvarez, J.C. Riquelme, Mining numeric association rules with genetic algorithms, in *Proceedings of the 5th International Conference on Artificial Neural Networks and Genetic Algorithms, ICANNGA 2001*, Taipei (2001), pp. 264–267

63. T.M. Mitchell, *Machine Learning*. McGraw Hill Series in Computer Science (McGraw-Hill, New York, 1997)

64. S. Moens, E. Aksehirli, B. Goethals, Frequent itemset mining for big data, in *Proceedings of the 2013 IEEE International Conference on Big Data*, Santa Clara, CA (2013), pp. 111–118

65. A.Y. Noaman, J.M. Luna, A.H.M. Ragab, S. Ventura, Recommending degree studies according to students' attitudes in high school by means of subgroup discovery. Int. J. Comput. Intell. Syst. **9**(6), 1101–1117 (2016)

66. P.K. Novak, N. Lavrač, G.I. Webb, Supervised descriptive rule discovery: a unifying survey of contrast set, emerging pattern and subgroup mining. J. Mach. Learn. Res. **10**, 377–403 (2009)

67. N. Ordoñez, C. Ezquerra, C. Santana, Constraining and summarizing association rules in medical data. Knowl. Inf. Syst. **9**, 259–283 (2006)

68. F. Padillo, J.M. Luna, S. Ventura, Mining perfectly rare itemsets on big data: an approach based on Apriori-inverse and mapreduce, in *Proceedings of the 16th International Conference on Intelligent Systems Design and Applications (ISDA 2016)*, Porto (2016), pp. 508–518

69. F. Padillo, J.M. Luna, S. Ventura, Subgroup discovery on big data: pruning the search space on exhaustive search algorithms, in *Proceedings of the 2016 IEEE International Conference on Big Data (BigData 2016)*, Washington DC (2016), pp. 1814–1823

70. F. Padillo, J.M. Luna, F. Herrera, S. Ventura, Mining association rules on big data through mapreduce genetic programming. Integr. Comput.-Aided Eng. **25**(1), 31–48 (2018)

71. J. Pei, G. Dong, W. Zou, J. Han, Mining condensed frequent-pattern bases. Knowl. Inf. Syst. **6**(5), 570–594 (2004)

72. C. Romero, S. Ventura, Educational data mining: a review of the state of the art. IEEE Trans. Syst. Man Cybern. Part C **40**(6), 601–618 (2010)

73. S. Ruggieri, D. Pedreschi, F. Turini, Data mining for discrimination discovery. ACM Trans. Knowl. Discov. Data (TKDD) **4**(2), 1–40 (2010)

74. D. Sánchez, J.M. Serrano, L. Cerda, M.A. Vila, Association rules applied to credit card fraud detection. Expert Syst. Appl. **36**, 3630–3640 (2008)

75. K.K.W. Siu, S.M. Butler, T. Beveridge, J.E. Gillam, C.J. Hall, A.H. Kaye, R.A. Lewis, K. Mannan, G. McLoughlin, S. Pearson, A.R. Round, E. Schultke, G.I. Webb, S.J. Wilkinson, Identifying markers of pathology in saxs data of malignant tissues of the brain. Nucl. Inst. Methods Phys. Res. A **548**, 140–146 (2005)

76. H.S. Song, J.K. Kimb, H.K. Soung, Mining the change of customer behavior in an internet shopping mall. Expert Syst. Appl. **21**(3), 157–168 (2001)

77. A. Soulet, B. Crémilleux, Adequate condensed representations of patterns. Data Min. Knowl. Disc. **17**(1), 94–110 (2008)

78. A. Soulet, B. Crmilleux, F. Rioult, Condensed representation of emerging patterns. in *Proceedings of the 8th Pacific-Asia Conference on Knowledge Discovery and Data Mining (PAKDD-04)*, Sydney (2004), pp. 127–132

79. P.N. Tan, M. Steinbach, V. Kumar, *Introduction to Data Mining* (Addison Wesley, Boston, 2005)

80. F. Thabtah, P. Cowling, Y. Peng, MMAC: a new multi-class, multi-label associative classification approach, in *Proceedings of the 4th IEEE International Conference on Data Mining (ICDM'04)*, Brighton (2004), pp. 217–224

81. F. Thabtah, P. Cowling, Y. Peng, MCAR: multi-class classification based on association rule approach, in *Proceedings of the 3rd IEEE International Conference on Computer Systems and Applications*, Cairo (2005), pp. 1–7

82. W. Ugarte, P. Boizumault, B. Crémilleux, A. Lepailleur, S. Loudni, M. Plantevit, C. Raïssi, A. Soulet, Skypattern mining: from pattern condensed representations to dynamic constraint satisfaction problems. Artif. Intell. **244**, 48–69 (2017)

83. S. Ventura, J.M. Luna, *Pattern Mining with Evolutionary Algorithms* (Springer International Publishing, Cham, 2016)

84. L. Wang, H. Zhao, G. Dong, J. Li, On the complexity of finding emerging patterns. Theor. Comput. Sci. **335**(1), 15–27 (2005)

85. G.I. Webb, Discovering associations with numeric variables, in *Proceedings of the Seventh ACM SIGKDD International Conference on Knowledge Discovery and Data Mining (KDD-2001)*, New York (2001), pp. 383–388

86. G.I. Webb, Discovering significant patterns. Mach. Learn. **71**(1), 131 (2008)

87. G.I. Webb, S.M. Butler, D.A. Newlands, On detecting differences between groups, in *Proceedings of the 9th ACM SIGKDD International Conference on Knowledge Discovery and Data Mining*, Washington, DC (2003), pp. 256–265

88. F. Wenbin, L. Mian, X. Xiangye, H. Bingsheng, L. Qiong, Frequent itemset mining on graphics processors, in *Proceedings of the 5th International Workshop on Data Management on New Hardware, DaMoN '09*, Providence, Rhode Island (2009), pp. 34–42

89. T.T. Wong, K.L. Tseng, Mining negative contrast sets from data with discrete attributes. Expert Syst. Appl. **29**(2), 401–407 (2005)

90. S. Wrobel, An algorithm for multi-relational discovery of subgroups, in *Proceedings of the 1st European Symposium on Principles of Data Mining and Knowledge Discovery, PKDD '97*, London (Springer, Berlin, 1997), pp. 78–87

91. X. Yin, J. Han, CPAR: classification based on predictive association rules, in *Proceedings of the 3rd SIAM International Conference on Data Mining, SDM 2003*, San Francisco, CA (2003), pp. 331–335
92. M.J. Zaki, Scalable algorithms for association mining. IEEE Trans. Knowl. Data Eng. **12**(3), 372–390 (2000)
93. C. Zhang, S. Zhang, *Association Rule Mining: Models and Algorithms* (Springer, Berlin, 2002)
94. W. Zhou, H. Wei, M.K. Mainali, K. Shimada, S. Mabu, K. Hirasawa, Class association rules mining with time series and its application to traffic load prediction, in *Proceedings of the 47th Annual Conference of the Society of Instrument and Control Engineers (SICE 2008)*, Tokyo (2008), pp. 1187–1192
95. F. Zhu, X. Yan, J. Han, P.S. Yu, H. Cheng, Mining colossal frequent patterns by core pattern fusion, in *Proceedings of the IEEE 23rd International Conference on Data Engineering, ICDE 2007*, Istanbul (IEEE, Piscataway, 2007), pp. 706–771

Chapter 2
Contrast Sets

Abstract Contrast set mining is one of the most important tasks in the supervised descriptive pattern mining field. It aims at finding patterns whose frequencies differ significantly among sets of data under contrast. This chapter introduces therefore the contrast set mining problem as well as similarities and differences with regard to related techniques. Then, the problem is formally described and the most widely used metrics in this task are mathematically defined. In this chapter, the most important approaches in this field are analysed, including STUCCO and CIGAR among others. Finally, some additional proposals in the field are also described.

2.1 Introduction

In any pattern mining task the key element is the pattern, which denotes important properties of data or an interesting subset of data with valuable common features [1]. When the subset to be analysed and described is previously identified or labeled, the problem is known under the term supervised descriptive pattern mining. In this framework, results are typically obtained in the form of rules where the antecedent states for the pattern or common features, whereas the consequent is responsible for denoting the target variable (class attribute).

Contrast set mining is one of the main areas of research in supervised descriptive pattern mining, also known as supervised descriptive rule discovery [14] due to results are generally given in form of rules. Contrasting may be performed on datasets satisfying statically defined conditions, aiming at discovering differences and similarities among datasets or given subsets of data. In the simplest way, such subsets were obtained by grouping records according to the class values of the general dataset, resulting in subsets with the same class value each. Sometimes, though, each subset (dataset under contrast) may contain different classes since the sets to compare were obtained by considering various conditions. For example, different diseases may produce different subsets (one per disease) to be analysed and each of these datasets may include a labeled variable with two classes, e.g. healthy and unhealthy.

© Springer Nature Switzerland AG 2018

S. Ventura, J. M. Luna, *Supervised Descriptive Pattern Mining*,

https://doi.org/10.1007/978-3-319-98140-6_2

Fig. 2.1 Two sets of shapes to contrast. The sets were split according to their color

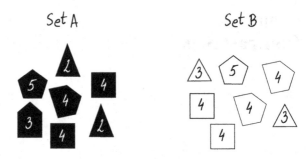

Contrast sets may be explained by means of a toy example (it is useless for additional purposes) including a set of shapes with different colors. Each figure or shape denotes a different record in data, including features such as whether it is regular or irregular (its sides and interior angles are all equal or not), its form (triangle, square or pentagon), its color (black or white) as well as the number of equal sides (labeled as a number inside). From this set of figures, two subsets were obtained to contrast according to its color (see Fig. 2.1). In this example, the shape cannot be a contrast feature since there is an equal number of triangles (2 in total), squares (2 in total) and pentagons (3 in total) in each of the sets. However, it is possible to assert that irregular triangles (not all its sides are equal) only appear in *Set A* (black shapes) so it is a really interesting contrast set for this example. Additionally, and considering the analysis of irregular polygons, it is obtained that there are 4 of them (57% of the figures) in black figures (*Set A*) and 2 of them (28% of the figures) in white figures (*Set B*). As a result, this may also be considered a contrast set since the frequency is almost the double from a set to another.

From the formal definition of contrast set mining provided by *Bay and Pazzani* [6], many different algorithms have been proposed by different researchers and they have been applied to different fields: retail sales data [18]; design of customized insurance programs [19]; identification of patterns in synchrotron x-ray data that distinguish tissue samples of different forms of cancerous tumor [16]; classify between two groups of brain ischaemia patients [12]. One of the most well-known approaches for mining contrast sets is STUCCO (Search and Testing for Understandable Consistent Contrasts) [6], which includes a pruning mechanism to discard and reduce the resulting number of contrast sets. In 2005, *Hilderman and Peckham* [11] proposed another algorithm, named CIGAR (ContrastIng Grouped Association Rules), with the aim of mining contrast sets which frequency is higher than a predefined threshold. Additionally, the contrast set mining problem has been solved by considering it as a subgroup discovery task, giving rise to the CSM-SD algorithm [13]. These and many other algorithms for mining contrast sets are studied in this chapter.

2.2 Task Definition

The contrast set mining problem was originally described [5] as an exploratory rule discovery technique for multinomial, grouped discrete attributes or variables. In a formal way, it was defined as the finding of patterns whose frequencies differ significantly among sets of data under contrast. Contrasting may be performed on datasets satisfying statically defined conditions or on subsets of a more general dataset.

In a forma way, let $I = \{i_1, i_2, ..., i_n\}$ be the set of items included in a dataset Ω that is represented in a binary tabular form (each single item may or may not appear in data) and let $T = \{t_1, t_2, ..., t_m\}$ be the collection of transactions or data records such as $\forall t \in T : t \subseteq I \in \Omega$. A contrast set is a pattern $P \subseteq I$ consisting of conjunctions of items from I, satisfying that the frequency of occurrence of P in a dataset is highly different to the frequency of the same pattern P in another dataset.

A contrast set may also be defined on a set of categorical attributes where $A = \{a_1, a_2, ..., a_m\}$ is the set of attributes in data, and $V_i = \{v_{i1}, v_{i2}, ..., v_{in}\}$ is the set of feasible discrete values associated with the attribute a_i. Here, a contrast set, which is denoted as a pattern P, is a conjunction of one or multiple attribute-value pairs in the form $P = \{(a_x, v_{xj}), ..., (a_y, v_{yk})\}$. Finally, even when the original definition was stated on discrete domains, contrast sets have been recently defined on continuous domains [15]. These quantitative contrast sets include attributes whose values correspond to a continuous range of numeric values. Now, each attribute a_i includes a range of values in the form $[v_{i_l}, v_{i_u}]$, v_{i_l} and v_{i_u} denoting the lower and upper bounds, respectively.

2.2.1 Quality Measures

When dealing with any pattern mining problem, as it is the case of supervised descriptive pattern mining, a large amount of different patterns may be obtained [17]. Given a dataset comprising n singletons or single items, i.e. $I = \{i_1, i_2, ..., i_n\}$, a maximum of $2^n - 1$ different patterns can be found, so any straightforward approach becomes extremely complex with the increasing number of singletons. It is therefore required to consider the right set of metrics to quantify how representative a specific pattern is for the dataset as well as its potential interest to the problem at hand.

Metrics in pattern mining are all based on the frequency (also known as support) of a pattern P, and it is generally defined as the number of transactions in which P appears, that is, $|\{\forall t \in T : P \subseteq t, t \subseteq I \in \Omega\}|$, $T = \{t_1, t_2, ..., t_m\}$ being the collection of transactions included in a dataset Ω. This frequency is usually defined in its relative form, stating for the percentage of transactions in which P appears among the whole set of transactions $|T|$, that is, $|\{\forall t \in T : P \subseteq t, t \subseteq I \in \Omega\}|/|T|$.

Table 2.1 Sample market basket datasets (Ω_1 left, Ω_2 right)

Items (dataset Ω_1)	Items (dataset Ω_2)
{Bread}	{Butter}
{Bread, Beer}	{Butter, Diaper}
{Bread, Beer, Diaper}	{Bread, Beer}
{Bread, Butter}	{Bread, Butter, Diaper}
{Bread, Beer, Butter, Diaper}	{Bread, Butter, Diaper}
{Beer}	{Bread}
{Beer, Diaper}	{Bread, Beer}
{Bread, Butter}	{Butter, Diaper}
{Bread, Beer, Diaper}	{Bread, Butter, Diaper}
{Bread}	{Butter}

For a better understanding, let us consider two sample datasets (see Table 2.1) and $support(P, \Omega_i)$ to denote the support of a pattern P on a dataset Ω_i. Taking the pattern $P = \{Bread, Beer\}$, then $support(P, \Omega_1) = 4/10 = 0.40$. A pattern P is defined as a closed pattern if for every super-pattern P', that is, $P \subset P'$, then $support(P, \Omega_i) < support(P', \Omega_i)$. Additionally, a pattern P is a minimal generator if for every pattern P'' such that $P'' \subset P$, then $support(P'', \Omega_i) > support(P, \Omega_i)$.

As it was previously introduced, when analysing the importance of a contrast set (represented as a pattern P) the key element is the difference in the frequencies (support values) of P on different datasets to be contrasted. These differences are quantified under the term support difference or support delta ($support_\delta$). For a matter of clarification, let us consider the simple case where two datasets Ω_1 and Ω_2 are taken (see Table 2.1)—these datasets can be considered as disjoint subsets of a more general dataset Ω. Here, the support difference [14] of a pattern P (candidate contrast set) is mathematically defined as the absolute differences between the support values of P in Ω_1 and Ω_2. In a mathematical way, it is defined as $support_\delta(P) = |support(P, \Omega_1) - support(P, \Omega_2)|$, where $support(P, \Omega_i)$ denotes the support of a pattern P on dataset Ω_i. For a better understanding, let us consider the pattern $P_x = \{Butter, Diaper\}$ having $support(P_x, \Omega_1) = 1/10 = 0.10$ and $support(P_x, \Omega_2) = 5/10 = 0.50$. Here, the support difference is calculated as $support_\delta(P_x) = |0.10 - 0.50| = 0.40$ so P_x may be defined as a good contrast set. On the contrary, the pattern $P_y = \{Bread, Butter\}$ has a support on Ω_1 that is equal to its support on Ω_2, that is, $support(P_y, \Omega_1) = support(P_y, \Omega_2) = 0.30$ so $support_\delta(P_y) = 0$. Hence, this pattern cannot be considered as a contrast set since its frequencies remain the same for both datasets.

In this problem it is also possible to apply a delta threshold σ_δ, determining a pattern P as a σ_δ-contrast pattern (or contrast set) if $support_\delta(P) \geq \sigma_\delta$. For example, taking a delta threshold $\sigma_\delta = 0.40$ and the datasets shown in Table 2.1, then the following contrast sets may be obtained: $P_1 = \{Beer\}$, $P_2 = \{Butter\}$,

Table 2.2 Sample market basket datasets (Ω_1, Ω_2, and Ω_3)

Items (dataset Ω_1)	Items (dataset Ω_2)	Items (dataset Ω_3)
{Bread}	{Butter}	{Bread, Butter}
{Bread, Beer}	{Butter, Diaper}	{Bread, Beer}
{Bread, Beer, Diaper}	{Bread, Beer}	{Bread, Beer, Diaper}
{Bread, Butter}	{Bread, Butter, Diaper}	{Bread, Butter}
{Bread, Beer, Butter, Diaper}	{Bread, Butter, Diaper}	{Beer, Butter, Diaper}
{Beer}	{Bread}	{Beer, Butter}
{Beer, Diaper}	{Bread, Beer}	{Beer, Diaper}
{Bread, Butter}	{Butter, Diaper}	{Bread, Butter}
{Bread, Beer, Diaper}	{Bread, Butter, Diaper}	{Bread, Beer, Diaper}
{Bread}	{Butter}	{Bread}

$P_3 = \{Beer, Diaper\}$, and $P_4 = \{Butter, Diaper\}$. All these contrast sets have a $support_\delta$ equal to 0.4 as it is demonstrated: $support_\delta(P_1) = |0.60 - 0.20| = 0.40$, $support_\delta(P_2) = |0.30 - 0.70| = 0.40$, $support_\delta(P_3) = |0.40 - 0.00| = 0.40$, and $support_\delta(P_4) = |0.10 - 0.50| = 0.40$.

Sometimes, however, it is also useful to apply a support threshold value α for P on Ω_i, in such a way that $support(P, \Omega_i) \geq \alpha$. This threshold enables the set of resulting contrast sets to be reduced. In this regard, let us suppose that the user is only interested in those sets that appear in at least 30% of the records, that is, $\alpha = 0.30$ in per unit basis. Considering the same $\sigma_\delta = 0.40$ previously used, then P_1 cannot be considered as a valid set since $support(P_1, \Omega_2) = 0.2$ and, therefore, it does not satisfy the minimum threshold. As a result, from the resulting set of contrast sets that satisfy the $support_\delta$ threshold, only P_2 also satisfies the α value.

Finally, contrast sets may consider more than two datasets, denoting a pattern P as a contrast set if and only if $\exists ij : max(|support(P, \Omega_i) - support(P, \Omega_j)|) \geq \sigma_\delta$. Let us consider now the two previous datasets Ω_1 and Ω_2 (see Table 2.1) and an additional one, denoted as Ω_3 (see Table 2.2). In this example, and considering the threshold value $\sigma_\delta = 0.40$, the following contrast sets are obtained: $P_1 = \{Beer\}$, $P_2 = \{Butter\}$, $P_3 = \{Beer, Diaper\}$, and $P_4 = \{Butter, Diaper\}$. As a matter of clarification, P_1 is a contrast set since $\exists ij : max(|support(P_1, \Omega_i) - support(P_1, \Omega_j)|) \geq 0.40$. In other words, $|support(P_1, \Omega_1) - support(P_1, \Omega_2)| = |0.60 - 0.20| = 0.40$, $|support(P_1, \Omega_1) - support(P_1, \Omega_3)| = |0.60 - 0.60| = 0.00$, and $|support(P_1, \Omega_2) - support(P_1, \Omega_3)| = |0.20 - 0.60| = 0.40$. Hence, P_1 satisfies the σ_δ threshold in two pairs of datasets and, therefore, there exists at least one combination of datasets in which P_1 is a contrast set. A minimum support threshold value α may also be considered so those datasets in which P_j is not frequent cannot be used to quantified the measure $support_\delta(P_j)$.

2.2.2   Tree Structures

Frequent pattern trees have been widely used in the pattern mining field for the discovery of high frequent itemsets in data [9]. Tree structures are demonstrated to be high efficient methods for mining patterns since they reduce the database scans by considering a compressed representation of the database and, therefore, allowing more data to be stored in main memory. This representation is based on a prefix-tree structure that represents each node by using a singleton and the frequency of a pattern is denoted by the path from the root to that node. As a result, the input database is compressed via sharing with common prefixes.

According to some authors [7], the use of a tree for mining contrast sets is evident given the popularity and success of frequent pattern trees for mining frequent itemsets. Nevertheless, the problem of mining frequent patterns is not essentially the same as the one of mining contrast sets so new challenges are considered. Now, two or more datasets are considered so the tree structure needs multiple counts to be kept in each node (one per dataset under contrast). Additionally, different methods for recursively exploring the tree are required, and two approaches are considered in this sense: ratio tree structure [4], and contrast pattern tree structure [8].

The ratio tree structure is quite similar to the frequent pattern tree structure described by *Han et al.* [9] except for the extra counts that each node includes (one per dataset). Here, each node is associated with a pattern and it stores the frequency of such a pattern in any of the datasets Ω_i. Analysing the tree structure, it is relatively easy to know whether a pattern is a contrast set since it is only required to check the difference in the counts (frequency of the pattern in Ω_i). At this point, and similarly to the frequent pattern tree structure, the global ordering of the items will have a huge impact in the final shape of the tree. This ordering will enable a minimization in the total number of nodes in the tree and will produce that more important nodes (those with higher $support_\delta$ values) will be higher up in the tree. According to *Dong et al.* [7], three different ordering can be considered as described below (results are exactly the same, and the main difference lies on the resulting tree structure).

2.2.2.1   Frequent Tree Ordering

A frequent tree ordering is the same process proposed by *Han et al.* [9] in which singletons or single items are ordered according to their frequency (descending probability). From this ordered list, items are inserted into the tree achieving a high compression of the data via many shared prefixes in the tree. In situations where more than a single datasets is considered (contrast set mining problem), the procedure is exactly the same as the one proposed by *Han et al.* [9] except for the probabilities which are calculated for each of the datasets in isolation. Given two datasets Ω_1 and Ω_2, the resulting probability is $p(\Omega_1 \cup \Omega_2)$.

In order to clarify this ordering strategy, let us consider the frequent tree ordering for two datasets (see Table 2.1). The frequency or support for each of the items (*Bread, Beer, Butter* and *Diaper*) is calculated as follows: $support(Bread, \Omega_1 \cup \Omega_2) = 14/20 = 0.70$; $support(Beer, \Omega_1 \cup \Omega_2) = 8/20 = 0.40$; $support(Butter, \Omega_1 \cup \Omega_2) = 10/20 = 0.50$; and $support(Diaper, \Omega_1 \cup \Omega_2) = 9/20 = 0.45$. The list of items is therefore sorted as *Bread* $\prec$ *Butter* $\prec$ *Diaper* $\prec$ *Beer*, and the resulting tree is formed as shown in Fig. 2.2.

Analysing the tree (see Fig. 2.2) and, more specifically, any resulting itemset occurring from the base to the root of such a tree, it is obtained that {*Beer, Bread*} is hardly a contrast set since it occurs one time in Ω_1 and two times in Ω_2. A better itemset is {*Beer, Diaper, Bread*}, which is satisfied two times in Ω_1 and it is not included in Ω_2. Traversing the tree, it is easy to check that {*Butter*}, {*Beer*}, {*Beer, Diaper*}, and {*Butter, Diaper*} are the most promising itemsets since their $support_\delta$ value is 0.4 in all of them.

This frequent tree ordering strategy can also be used to mine contrast sets on more than two datasets (see Table 2.2). In this example, the frequency or support for each of the items (*Bread, Beer, Butter* and *Diaper*) is obtained as follows: $support(Bread, \Omega_1 \cup \Omega_2 \cup \Omega_3) = 21/30 = 0.70$; $support(Beer, \Omega_1 \cup \Omega_2 \cup \Omega_3) = 14/30 = 0.47$; $support(Butter, \Omega_1 \cup \Omega_2 \cup \Omega_3) = 15/30 = 0.50$; and $support(Diaper, \Omega_1 \cup \Omega_2 \cup \Omega_3) = 13/30 = 0.43$. The list of items is therefore sorted as *Bread* $\prec$ *Butter* $\prec$ *Beer* $\prec$ *Diaper*, and the resulting tree is formed as shown in Fig. 2.3. Taking itemsets from the bottom to the root, it is obtained that {*Bread, Beer, Diaper*} is a contrast set that appears two times in Ω_1 and Ω_3, but it does not appear in Ω_2. From the resulting tree (see Fig. 2.3), it is obtained that the best contrast sets are the following: {*Beer*}, {*Butter*}, {*Beer, Diaper*}, and {*Butter, Diaper*}. In all of them, it is satisfied that $max(|support(P, \Omega_i) - support(P, \Omega_j)|) = 0.40$.

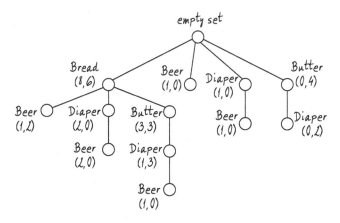

Fig. 2.2 Frequent tree ordering for datasets Ω_1 and Ω_2. Resulting ordering is *Bread* $\prec$ *Butter* $\prec$ *Diaper* $\prec$ *Beer*

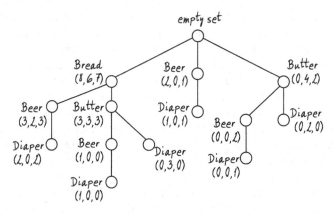

Fig. 2.3 Frequent tree ordering for datasets Ω_1, Ω_2 and Ω_3. Resulting ordering is *Bread* $\prec$ *Butter* $\prec$ *Beer* $\prec$ *Diaper*

2.2.2.2 Difference Ordering

Taking two datasets Ω_1 and Ω_2, and the frequency or support of an item P for such datasets as $support(P, \Omega_1)$ and $support(P, \Omega_2)$; items are sorted according to $support_\delta(P) = |support(P, \Omega_1) - support(P, \Omega_2)|$. This ordering aims to allocate better contrast sets (those with a higher $support_\delta$ value) higher up in the tree. The aim is to discover such contrast sets earlier in the mining process and, therefore, a more effective prunning of the bottom nodes in the tree can be performed.

As an example, let us consider this ordering strategy for two sample datasets (see Table 2.1). Here, the frequency for each single item (*Bread*, *Beer*, *Butter* and *Diaper*) is used to calculate the $support_\delta$ metric. As it was previously defined, given an itemset or pattern P in two datasets (Ω_1 and Ω_2), $support_\delta(P) = |support(P, \Omega_1) - support(P, \Omega_2)|$. Thus, taking the aforementioned items, the following values are obtained: $support_\delta(Bread) = |0.8 - 0.6| = 0.2$; $support_\delta(Beer) = |0.6 - 0.2| = 0.4$; $support_\delta(Butter) = |0.3 - 0.7| = 0.4$; and $support_\delta(Diaper) = |0.4 - 0.5| = 0.1$. The list of items is therefore sorted as *Beer* $\prec$ *Butter* $\prec$ *Bread* $\prec$ *Diaper*, and the resulting tree is formed as shown in Fig. 2.4. Analysing this tree, it is easy to check that *Beer* and *Butter* are really promising sets, and when these items are combined with *Diaper* high interesting sets are also found. Here, the contrast sets are exactly the same as the ones obtained with the frequent tree ordering strategy, and the main difference lies on the way in which the tree is formed.

When dealing with more than two datasets (see Table 2.2), items are sorted according to $support_\delta$ for all the pairs of datasets so the maximum value is considered. In a formal way, an item (or itemset) P is sorted according to the value $\forall ij : max(|support(P, \Omega_i) - support(P, \Omega_j)|)$. Given the set of

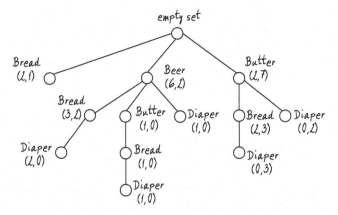

Fig. 2.4 Difference tree ordering for datasets Ω_1 and Ω_2. Resulting ordering is *Beer* ≺ *Butter* ≺ *Bread* ≺ *Diaper*

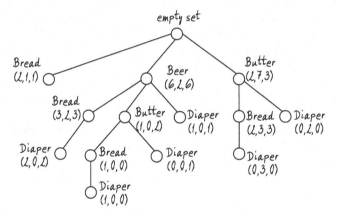

Fig. 2.5 Difference tree ordering for datasets Ω_1, Ω_2 and Ω_3. Resulting ordering is *Beer* ≺ *Butter* ≺ *Bread* ≺ *Diaper*

items (*Bread*, *Beer*, *Butter* and *Diaper*), the following values are obtained for each of them: $max(|support(Bread, \Omega_i) - support(Bread, \Omega_j)|) = 0.2$; $max(|support(Beer, \Omega_i) - support(Beer, \Omega_j)|) = 0.4$; $max(|support(Butter, \Omega_i) - support(Butter, \Omega_j)|) = 0.4$; and $max(|support(Diaper, \Omega_i) - support(Diaper, \Omega_j)|) = 0.1$. The list of items is therefore sorted as *Beer* ≺ *Butter* ≺ *Bread* ≺ *Diaper*, and the resulting tree is formed as shown in Fig. 2.5.

2.2.2.3 Hybrid Ordering

The hybrid ordering performs a combination of the two previous ordering strategies, that is, difference ordering and frequent tree ordering. In this process, both the

difference of supports and the frequency for each items are calculated. Then, a
user specified percentage is provided so items within such percentage are chosen
according to the difference ordering, whereas the rest of items are selected according
to their frequency. As a result, this process creates trees that share features of the
two previous ordering procedures.

In order to clarify this sorting strategy, let us consider both the frequent tree
ordering and the difference ordering strategies for two datasets (see Table 2.1).
According to the frequent tree ordering, the list of items is sorted as *Bread* ≺
Butter ≺ *Diaper* ≺ *Beer* (see Sect. 2.2.2.1), whereas the same list is sorted
as *Beer* ≺ *Butter* ≺ *Bread* ≺ *Diaper* when the difference sorting strategy
is considered (see Sect. 2.2.2.2). In this example, let us take the first 50% of the
items according to the frequent tree ordering, and the remaining 50% based on the
difference sorting strategy. Thus, the resulting arranged list is *Bread* ≺ *Butter* ≺
Beer ≺ *Diaper*, and the resulting tree is formed as shown in Fig. 2.6.

This hybrid ordering strategy can also be applied to more than three datasets
(see Table 2.2). For three datasets and considering the frequent tree ordering, the
list of items is sorted as *Bread* ≺ *Butter* ≺ *Beer* ≺ *Diaper* (see Sect. 2.2.2.1),
whereas the same list is sorted as *Beer* ≺ *Butter* ≺ *Bread* ≺ *Diaper* when the
difference sorting strategy is considered (see Sect. 2.2.2.2). Similarly to the previous
example for two datasets, in this example, the first 50% of the items are taken from
the list sorted by the frequent tree ordering, whereas the remaining 50% are given
by the difference sorting strategy. As a result, the list of items is sorted as *Bread* ≺
Butter ≺ *Beer* ≺ *Diaper*, forming the tree shown in Fig. 2.7.

Any of the trees can be traversed to form itemsets that represent contrast sets. It
is important to remark that the sorting strategy does not affect the final results (set
of contrast sets) but the form of the tree. Thus, any of the three described strategies
provide the same contrast sets. Four are the most promising contrast sets when two

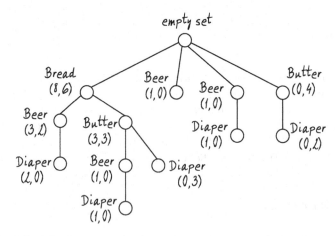

Fig. 2.6 Hybrid ordering for datasets Ω_1 and Ω_2. Resulting ordering is *Bread* ≺ *Butter* ≺
Beer ≺ *Diaper*

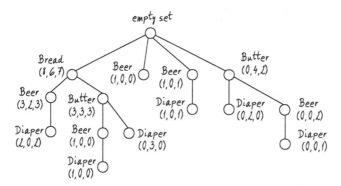

Fig. 2.7 Hybrid ordering for datasets Ω_1, Ω_2 and Ω_3. Resulting ordering is *Bread* $\prec$ *Butter* $\prec$ *Beer* $\prec$ *Diaper*

datasets (Ω_1 and Ω_2) or three datasets (Ω_1, Ω_2 and Ω_3) are considered: {*Butter*}, {*Beer*}, {*Beer, Diaper*}, and {*Butter, Diaper*}.

2.3 Algorithms for Mining Contrast Sets

Contrast set mining was defined in 1999 by *Bay and Pazzani* [5] as a data analysis task to understand differences between several contrasting groups. It enables to know, for example, which symptoms differentiate similar diseases; differences between customers that spend less money and those who spend more in a specific item; or even the difference between people holding PhD degrees and people holding Bachelor degrees. Contrast sets are conjunctions of attribute-value pairs whose distribution generally differ among groups. Since its definition, many different algorithms have been proposed in literature and the most important ones are described in this section.

2.3.1 STUCCO

The STUCCO (Search and Testing for Understandable Consistent COntrasts) algorithm is one of the first algorithms proposed for mining contrast sets. It was defined by *Bay and Pazzani* [5] with the aim of finding contrasts sets that make one group different than another. This difference is measured as the support value of each contrast set across groups, which should be greater than a minimum value σ_δ (all these concepts were described in Sect. 2.2.1). STUCCO considers the problem of mining contrast sets as a tree search problem. Similarly to Apriori [2] the search space is represented in a Hasse diagram where the root node represents an empty contrast set, whereas each children of a node is generated by adding an item to

---

Algorithm 2.1 Pseudo-code of the STUCCO algorithm

Require: $\Omega, \sigma_\delta, \alpha$ ▷ dataset, minimum difference value, and alpha value
Ensure: S
1: $C \leftarrow \emptyset$ ▷ set of candidates
2: $S \leftarrow \emptyset$ ▷ set of solutions
3: $k \leftarrow 1$ ▷ length of the candidates
4: $C \leftarrow items \in \Omega$ of size k ▷ set of single items or singletons within Ω
5: **while** $C \neq \emptyset$ **do**
6:      **for all** $c \in C$ **do**
7:          compute support c ▷ support of c on each group $\Omega_i \in \Omega$
8:          **if** $\exists ij : max(|support(c, \Omega_i) - support(c, \Omega_j)|) \geq \sigma_\delta$ **then**
9:              **if** $\chi^2(c) \geq \alpha$ **then**
10:                 $S \leftarrow c$ ▷ c is a valid contrast-set
11:             **end if**
12:         **end if**
13:     **end for**
14:     $k \leftarrow k + 1$
15:     $C \leftarrow$ generate items of size k from S
16: **end while**
17: **return** S

---

the parent. A node in the tree is pruned if it is neither significant (considering an statistical test) nor large (the difference between supports for each group is greater than a minimum σ_δ value).

STUCCO is an iterative algorithm (see Algorithm 2.1) that works in a breadth-first manner by producing any feasible itemset of size k. At the beginning, the algorithm keeps in C all the single items (also known as singletons) from dataset Ω (see line 4, Algorithm 2.1). Then, STUCCO iterates till the set of candidate itemsets is empty (see lines 5 to 16, Algorithm 2.1). In the k-th iteration, any itemset of size k is analysed by computing its support on each of the groups as well as the chi-square value (χ^2). A specific itemset $c \in C$ will be considered as a valid contrast-set if and only if $\exists ij : max(|support(c, \Omega_i) - support(c, \Omega_j)|) \geq \sigma_\delta$ and $\chi^2(c) \geq \alpha$. Both σ_δ and α are values predefined by the user. Finally, the algorithm returns the set of solutions once all the levels in the search space are reached or no more candidate itemsets can be produced (see line 17, Algorithm 2.1).

Let us consider the sample market basket datasets (Ω_1, Ω_2, and Ω_3) shown in Table 2.2 and the Hasse diagram illustrated in Fig. 2.8. STUCCO searches the tree (the aforementioned Hasse diagram) in a breadth-first, levelwise manner. For each specific level (from the root to the leaves of the tree), the database is analysed and the support of each node (itemset or candidate contrast-set) for each group is calculated (see Fig. 2.9). Taking a σ_δ value equal to 0.4, it is possible to prune the nodes $Bread$ and $Diaper$ since $\nexists ij : max(|support(Bread, \Omega_i) - support(Bread, \Omega_j)|) \geq 0.40$ nor $\nexists ij : max(|support(Diaper, \Omega_i) - support(Diaper, \Omega_j)|) \geq 0.40$. Then, an statistical test (chi-square with a threshold $\alpha = 0.05$) is performed on those nodes that satisfy the minimum σ_δ value, that is, $Beer$ and $Butter$. Here, the null hypothesis is "The support for the contrast-set is the same across all groups".

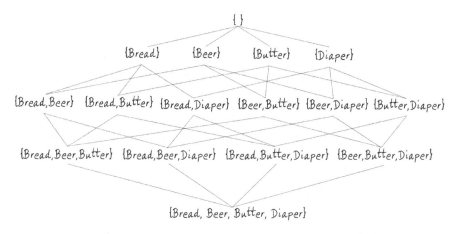

Fig. 2.8 Hasse diagram for four different items included in datasets Ω_1, Ω_2 and Ω_3

Fig. 2.9 Support values
(considering the datasets Ω_1,
Ω_2 and Ω_3) for each node in
the first level of the tree

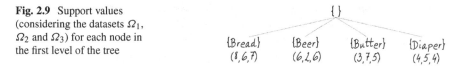

The chi-square test is computed as follows on a $2 \times nColumns$ contingency table
where the rows represent the truth (and untruth) of the contrast set (r takes the
value 2), and the columns indicate the group for a total of $nColumns$: $\chi^2 = \sum_{i=1}^{r} \sum_{j=1}^{nColumns} \frac{(O_{ij}-E_{ij})^2}{E_{ij}}$. In this equation, O_{ij} is the observed frequency count in
cell ij, and E_{ij} is the expected frequency count in cell ij given independence of the
row and column variables and is calculated as $E_{ij} = \sum_{i=1}^{r} O_{ij} \sum_{j=1}^{nColumns} O_{ij}/N$,
N being the total number of observations. The number of degrees of freedom for χ^2
is given by $(r-1) \times (nColumns - 1)$. Hence, taking the *Beer* node, $\chi^2 = 4.2857$
and the p-value is 0.117319. Given the *Butter* node, $\chi^2 = 3.2$ and the p-value
is 0.201897. None of these two nodes are significant for the STUCCO algorithms
and, therefore, there is no interesting contrast-sets (all single nodes are pruned) on
datasets Ω_1, Ω_2, and Ω_3.

Once a set of solutions is returned by STUCCO, it is also possible to reduce
the number of solutions by keeping those that are really surprising. A contrast set
is considered to be surprising if its support is different from what is expected. Let
us consider two contrast sets P_1 and P_2 defined on a dataset Ω. If we know that
$support(P_1, \Omega) = 0.2$ and $support(P_2, \Omega) = 0.7$ then, under the independence of
P_1 and P_2 it is expected that $support_{expected}(P_1 \wedge P_2, \Omega) = 0.2 \times 0.7 = 0.14$. Here,
if the actual value is different, then the result is surprising, that is, $support(P_1 \wedge P_2, \Omega) = 0.2 \neq support_{expected}(P_1 \wedge P_2, \Omega) = 0.14$.

2.3.2 CIGAR

CIGAR (ContrastIng Grouped Association Rules) was proposed by *Hilderman et al.* [11] in 2005. This algorithm is really similar to STUCCO [5] and the main difference lies on the contingency tables used in both cases. Whereas STUCCO considers the analysis of $2 \times nColumns$ contingency tables to analyse differences among $nColumns$ groups, CIGAR breaks the $2 \times nColumns$ contingency tables down into a series of 2×2 contingency tables to try to explain where the differences actually occur.

CIGAR is similar to STUCCO as shown in Algorithm 2.2 with the difference that it is more restrictive for a candidate set to become a contrast set. Thus, even when CIGAR considers the general definition that determines a contrast set P as a candidate if $\exists ij : max(|support(P, \Omega_i) - support(P, \Omega_j)|) \geq \sigma_\delta$, three additional constraints are also used: $support(P, \Omega_i) \geq \beta$, $correlation(P, \Omega_i) \geq \lambda$, and $|correlation(P, \Omega_i) - correlation(child(P), \Omega_i)| \geq \gamma$. Here, β is the user-defined minimum support threshold, λ is the user-defined minimum correlation threshold, and γ is the user-defined minimum correlation difference. These three additional constraints are useful to identify frequent, strong and non-spurious sets. Considering the condition $support(P, \Omega_i) \geq \beta$ is used to identify frequent sets and to avoid outliers since they can dramatically affect the correlation value. Condition

---

Algorithm 2.2 Pseudo-code of the CIGAR algorithm

Require: $\Omega, \sigma_\delta, \alpha, \beta, \gamma, \delta$ ▷ dataset and minimum threshold values
Ensure: S
 1: $C \leftarrow \emptyset$ ▷ set of candidates
 2: $S \leftarrow \emptyset$ ▷ set of solutions
 3: $k \leftarrow 1$ ▷ length of the candidates
 4: $C \leftarrow items \in \Omega$ of size k ▷ set of single items or singletons within Ω
 5: **while** $C \neq \emptyset$ **do**
 6:      **for all** $c \in C$ **do**
 7:          compute support c ▷ support of c on each group $\Omega_i \in \Omega$
 8:          **if** $\exists ij : max(|support(c, \Omega_i) - support(c, \Omega_j)|) \geq \sigma_\delta$ **then**
 9:             **if** $\chi^2(c) \geq \alpha$ **then**
10:                **if** $\forall \Omega_i \in \Omega : support(c, \Omega_i) \geq \beta$ **then**
11:                    **if** $\forall \Omega_i \in \Omega : correlation(c, \Omega_i) \geq \lambda$ **then**
12:                        **if** $\forall \Omega_i \in \Omega : |correlation(c, \Omega_i) - correlation(child(c), \Omega_i)| \geq \gamma$ **then**
13:                          $S \leftarrow c$ ▷ c is a valid contrast-set
14:                      **end if**
15:                  **end if**
16:                **end if**
17:             **end if**
18:          **end if**
19:      **end for**
20:      $k \leftarrow k + 1$
21:      $C \leftarrow$ generate items of size k from S
22: **end while**
23: **return** S

Table 2.3 Generic contingency table of size 2×2

	Ω_1	Ω_2	ΣRow
Contrast set	O_{11}	O_{12}	$O_{11} + O_{12}$
$\neg$ Contrast set	O_{21}	O_{22}	$O_{21} + O_{22}$
$\Sigma Column$	$O_{11} + O_{21}$	$O_{12} + O_{22}$	$O_{11} + O_{12} + O_{21} + O_{22}$

$correlation(P, \Omega_i) \geq \lambda$ is used to identify strong sets, measuring the strength of any linear relationship between the contrast set and group membership. Finally, $|correlation(P, \Omega_i) - correlation(child(P), \Omega_i)| \geq \gamma$ is used to avoid spurious sets that should be pruned from the search space.

In order to determine whether a candidate set c is correlated, the Phi correlation coefficient is considered. This coefficient is a measure of the degree of association between two dichotomous variables contained in a 2×2 contingency table (see Table 2.3), and it compares the diagonal cells to the off-diagonal cells. Based on this coefficient, the variables are considered as positively associated (a value of 1 is obtained) if data is concentrated along the diagonal, whereas these variables are defined as negative associated (a value of -1 is obtained) if data is concentrated off the diagonal. A zero value represents no relationship between the variables. The Phi correlation coefficient is formally defined as $\phi = (O_{11}O_{22} - O_{12}O_{21})/\sqrt{(O_{11} + O_{21})(O_{12} + O_{22})(O_{11} + O_{12})(O_{21} + O_{22})}$. According to authors [11], since χ^2 value is anyway required in CIGAR, the Phi correlation may be expressed as $r = \sqrt{\chi^2/N}$, N being $O_{11} + O_{12} + O_{21} + O_{22}$.

As a matter of clarification, let us take the sample market basket dataset shown in Table 2.2, which comprises three different groups (Ω_1, Ω_2 and Ω_3), and the candidate pattern *Beer*. For this sample pattern (candidate contrast set) three different contingency tables of size 2×2 are obtained (see Table 2.4). In the first contingency table (groups Ω_1 and Ω_2), the correlation coefficient is computed as $r = \sqrt{3.3333/20} = 0.4082$ which is exactly the same value obtained by Phi as $\phi = 40/\sqrt{9,600} = 0.4082$. This value is $r = \sqrt{0/20} = 0$ for the second contingency table (groups Ω_1 and Ω_3). Finally, the value is $r = \sqrt{3.3333/20} = 0.4082$ for the third contingency table (groups Ω_2 and Ω_3). As a result, there is a weak difference between Ω_1 and Ω_2 as well as between Ω_2 and Ω_3. On the contrary, there is no difference between Ω_1 and Ω_3.

The main difference between STUCCO [5] and CIGAR [11] is therefore based on the contingency tables. Whereas the former uses a $2 \times nGroups$ contingency table, CIGAR uses a series of 2×2 contingency tables, one for each possible combination of group pairs in $nGroups$. Hence, CIGAR enables to know not only whether there are significant differences between some groups, but also which groups are involved in it. On the contrary, STUCCO provides a general knowledge about whether exist differences among a set of groups. Of course, for those analyses that include only two groups, the knowledge provided by these two algorithms is the same—the number of contrast sets may vary due to different constraints used in each of these algorithms.

Table 2.4 Contingency
tables of size 2×2 obtained
for variable *Beer* on three
different groups (Ω_1, Ω_2
and Ω_3)

	Ω_1	Ω_2	ΣRow
{*Beer*}	6	2	8
¬{*Beer*}	4	8	12
$\Sigma Column$	10	10	20

	Ω_1	Ω_3	ΣRow
{*Beer*}	6	6	12
¬{*Beer*}	4	4	8
$\Sigma Column$	10	10	20

	Ω_2	Ω_3	ΣRow
{*Beer*}	2	6	8
¬{*Beer*}	8	4	12
$\Sigma Column$	10	10	20

2.3.3   CSM-SD

CSM-SD (Contrast Set Mining—Subgroup Discovery) [13] was proposed in 2009
by considering that subgroup discovery [14] techniques can be used to mine
interesting contrast sets. According to authors, the main conceptual mismatch
between contrast set mining and subgroup discovery is the input to the algorithm.
While in contrast set mining considers the contrasting groups as input of the
algorithm, subgroup discovery obtains subgroups as the output of the algorithm.
Additionally, all the contrasting groups are equally important in contrast set mining,
but this assertion is not completely true in subgroup discovery.

To understand the compatibility of contrast set mining and subgroup discovery,
let us consider a special case of contrast set mining where only two contrasting
groups Ω_1 and Ω_2 are considered. For this problem, two different subgroup
discovery tasks can be applied, that is, one considering $C = \Omega_1$ and $\overline{C} = \Omega_2$, and
other one considering $C = \Omega_2$ and $\overline{C} = \Omega_1$. According to authors [13], since this
translation is possible for a two-group contrast set mining task, it is also possible for
a general contrast set mining task involving n contrasting groups. At this point, it is
important to remark that there are two feasible ways of applying these problems to
multi-class learning problems: (1) rules characterize one class compared to the rest
of the data (the standard one-versus-all setting); (2) rules that discriminate between
all pairs of classes.

CSM-SD is a simple algorithm (see Algorithm 2.3) where each pair of contrasting groups are considered as classes to be analysed by a subgroup discovery
algorithm. Here, the most important issue is that support difference used in
contrast set mining is compatible with weighted relative accuracy used in subgroup
discovery. Two quality measures f_1 and f_2 are considered as compatible if its values
follow the same behaviour on any pair of solutions, that is, $\forall R_i, R_j : f_1(R_i) >
f_1(R_j) \iff f_2(R_i) > f_2(R_j)$.

Support difference was previously defined in Sect. 2.2.1 as the absolute differences between support values of a pattern P (a candidate contrast set) in Ω_1 and
Ω_2. In a mathematical way, it was defined as $support_\delta(P) = |support(P, \Omega_1) -$

Algorithm 2.3 Pseudo-code of the CSM-SD algorithm

Require: Ω
Ensure: S
1: $C \leftarrow \{\Omega_1, \Omega_2, ..., \Omega_n\} \in \Omega$ ▷ set of groups in Ω to be analysed
2: **for all** $c_i \in C$ **do**
3:        **for all** $c_j \in C : i < j$ **do**
4:              apply a Subgroup Discovery algorithm for classes c_i and c_j
5:        **end for**
6: **end for**
7: **return** S

$support(P, \Omega_2)|$, where $support(P, \Omega_i)$ denotes the support of a pattern P on dataset Ω_i. In subgroup discovery, however, measures are based on the covering property so the true positive rate $TP(P, \Omega_i)$ is defined as the percentage of positive examples correctly classified as positive by P, and the false positive rate $FP(P, \Omega_i)$ is defined as the percentage of negative examples incorrectly classified as positive by P. Thus, $support_\delta(P)$ can also be denoted as $support_\delta(P) = TP(P, \Omega_1) - TP(P, \Omega_2) = TP(P, \Omega_1) - FP(P, \Omega_1)$. Finally, as for the weighted relative accuracy [10], which is widely used in subgroup discovery, authors of CSM-SD[13] demonstrated that once this metric is maximized, then $TP(P) - FP(P)$ is also maximized, which actually is the support difference $support_\delta(P)$. It demonstrated that the use of a subgroup discovery algorithm where weighted relative accuracy is maximized is similar to look for contrasting sets.

2.3.4   Additional Approaches

In 2005, *Wong et al.* [19] proposed a an approach for mining negative contrast sets. These show a significant difference between the existence of some characteristics and the nonexistence of some other characteristics for various groups in data. Given a pattern P (a feasible candidate contrast set) comprising a set of items $i_1, i_2, ..., i_n$, it is denoted $\neg i_j$ as the negation of the item $i_j \in P$. If any single item i_j does not occur, then it is said that $\neg i_j$ occurs. In the proposed approach [19], an itemset or candidate contrast set P is a negative itemset if some bipartition $\{X, Y\} \in P$ satisfies the following conditions:

- $support(X, \Omega_i) \geq \alpha$ and $support(Y, \Omega_i) \geq \alpha$, for a minimum support threshold value α.
- Any of the following is satisfied: $support(X \cup \neg Y, \Omega_i) \geq \alpha$; $support(\neg X \cup Y, \Omega_i) \geq \alpha$; $support(\neg X \cup \neg Y, \Omega_i) \geq \alpha$.
- Giving a minimum interest value β, any condition is satisfied: $interest(X \cup \neg Y, \Omega_i) = |support(X \cup \neg Y, \Omega_i) - support(X, \Omega_i) \times support(\neg Y, \Omega_i)| \geq \beta$; or $interest(\neg X \cup Y, \Omega_i) = |support(\neg X \cup Y, \Omega_i) - support(\neg X, \Omega_i) \times support(Y, \Omega_i)| \geq \beta$; or $interest(\neg X \cup \neg Y, \Omega_i) = |support(\neg X \cup \neg Y, \Omega_i) - support(\neg X, \Omega_i) \times support(\neg Y, \Omega_i)| \geq \beta$.

Another approach for mining contrast sets was proposed in 2007 by *Hilderman et al.* [15]. Contrast set mining techniques are generally focused on categorial data since generating contrast sets containing continuous-valued attributes is not straightforward. In this regard, authors proposed a modified equal-width binning interval approach to discretize continuous-valued attributes. This algorithm, known as Gen_QCSets, requires a set of approximate interval widths associated to each attribute. In each iteration, a different attribute is discretized according to the predefined interval widths, and each of these intervals are analyzed with regard to each of the groups Ω_i and Ω_j. Then, in order to be able to apply a series of t-test and z-test, it is required that at least two examples are included in each interval. To guarantee this issue, the upper bound of the interval is repeatedly increased by a predefined factor till at least two examples are included (satisfying the requirements). At this point, it is interesting to note that, when the last interval does not include at least two examples, there is no possibility to be expanded and, therefore, the previous interval is expanded to include it. In other words, the last two intervals from a new one.

In 2016, contrast set mining was considered as a Big Data problem [3]. This task, which may be easily carried out in small datasets, becomes prohibitively expensive due to its exponential complexity specially with Big Data. In this regard, authors proposed a greedy algorithm, called DisCoSet, to incrementally find a minimum set of local features that best distinguishes a group (defined as a class) from other groups (classes) without resorting to discretization. Unlike previous works that generate a large amount of candidate contrast sets to reduce it in a posteriori phase (keeping only the interesting ones), the DisCoSet prunes the search space while searching for the contrast sets. DisCoSet selects only features that improve the discriminative power of the contrast set of a group.

References

1. C.C. Aggarwal, J. Han, *Frequent Pattern Mining* (Springer International Publishing, Berlin, 2014)
2. R. Agrawal, T. Imielinski, A.N. Swami, Mining association rules between sets of items in large databases, in *Proceedings of the 1993 ACM SIGMOD International Conference on Management of Data*, SIGMOD Conference '93, pp. 207–216, Washington, DC (1993)
3. Z. Al Aghbari, I.N. Junejo, Discovering contrast sets for efficient classification of big data, in *Proceedings of the 2nd International Conference on Open and Big Data (OBD 2016)*, pp. 45–51, Vienna, August 2016
4. J. Bailey, T. Manoukian, K. Ramamohanarao, Fast algorithms for mining emerging patterns, in *Proceedings of the 6th European Conference on Principles of Data Mining and Knowledge Discovery*, PKDD '02, pp. 39–50, London (2002)
5. S.D. Bay, M.J. Pazzani, Detecting change in categorical data: mining contrast sets, in *Proceedings of the 5th ACM SIGKDD International Conference on Knowledge Discovery and Data Mining (KDD-99)*, pp. 302–306, New York (1999)
6. S.D. Bay, M.J. Pazzani, Detecting group differences: mining contrast sets. Data Min. Knowl. Disc. **5**(3), 213–246 (2001)

7. G. Dong, J. Bailey (eds.), *Contrast Data Mining: Concepts, Algorithms, and Applications* (CRC Press, Boca Raton, 2013)
8. H. Fan, K. Ramamohanarao, Fast discovery and the generalization of strong jumping emerging patterns for building compact and accurate classifiers. IEEE Trans. Knowl. Data Eng. **18**(6), 721–737 (2006)
9. J. Han, J. Pei, Y. Yin, R. Mao, Mining frequent patterns without candidate generation: a frequent-pattern tree approach. Data Min. Knowl. Disc. **8**, 53–87 (2004)
10. F. Herrera, C.J. Carmona, P. González, M.J. del Jesus, An overview on subgroup discovery: foundations and applications. Knowl. Inf. Syst. **29**(3), 495–525 (2011)
11. R.J. Hilderman, T. Peckham, A statistically sound alternative approach to mining contrast sets, in *Proceedings of the 4th Australasian Data Mining Conference (AusDM)*, pp. 157–172, Sydney, December 2005
12. N. Lavrac, P. Kralj, D. Gamberger, A. Krstacic, Supporting factors to improve the explanatory potential of contrast set mining: analyzing brain ischaemia data, in *Proceedings of the 11th Mediterranean Conference on Medical and Biological Engineering and Computing (MEDICON-07)*, pp. 157–161, Ljubljana, June 2007
13. P.K. Novak, N. Lavrač, D. Gamberger, A. Krstačic, CSM-SD: methodology for contrast set mining through subgroup discovery. J. Biomed. Inform. **42**(1), 113–122 (2009)
14. P.K. Novak, N. Lavrač, G.I. Webb, Supervised descriptive rule discovery: a unifying survey of contrast set, emerging pattern and subgroup mining. J. Mach. Learn. Res. **10**, 377–403 (2009)
15. M. Simeon, R.J. Hilderman, Exploratory quantitative contrast set mining: A discretization approach, in *Proceedings of the 19th IEEE International Conference on Tools with Artificial Intelligence (IEEE ICTAI'07)*, pp. 124–131, Patras, October 2007
16. K.K.W. Siu, S.M. Butler, T. Beveridge, J.E. Gillam, C.J. Hall, A.H. Kaye, R.A. Lewis, K. Mannan, G. McLoughlin, S. Pearson, A.R. Round, E. Schultke, G.I. Webb, S.J. Wilkinson, Identifying markers of pathology in SAXS data of malignant tissues of the brain. Nucl. Inst. Methods Phys. Res. A **548**, 140–146 (2005)
17. S. Ventura, J.M. Luna, *Pattern Mining with Evolutionary Algorithms* (Springer International Publishing, Berlin, 2016)
18. G.I. Webb, S.M. Butler, D.A. Newlands, On detecting differences between groups, in *Proceedings of the 9th ACM SIGKDD International Conference on Knowledge Discovery and Data Mining*, pp. 256–265, Washington, DC, August 24 (2003)
19. T.T. Wong, K.L. Tseng, Mining negative contrast sets from data with discrete attributes. Expert Syst. Appl. **29**(2), 401–407 (2005)

Chapter 3
Emerging Patterns

Abstract Emerging pattern mining is a well-known task in the supervised descriptive pattern mining field. This task aims at discovering emerging trends amongst timestamped datasets or extracting patterns that denote a clear difference between two disjoint features. This chapter introduces therefore the emerging pattern mining problem and describes the main differences with regard to contrast set mining. Then, the task is formally described as well as the existing types of emerging patterns. Finally, the most important approaches in this field are analysed, which are categorized into four different types (border-based, constraint-based, tree-based, and evolutionary fuzzy system-based).

3.1 Introduction

Emerging pattern mining was first proposed by *Dong et al.* [6] in 1999 as an important data mining task that aims at discovering discriminative patterns which can describe an emerging behaviour with respect to a property of interest. This emerging behaviour is measured as a significant increment in the support (frequency of occurrence or number of times the patterns is satisfied) from one dataset to another. This task therefore belongs to the pattern mining field since it looks for patterns (set of items) that denotes important properties of data [27]. More specifically, the emerging pattern mining problem is categorized under the term supervised descriptive pattern mining since it describes subsets that are previously identified or labeled.

This supervised descriptive pattern mining task is highly related to the one of mining contrast sets (see Chap. 2). In fact, some authors [5] considered the two concepts (contraset sets and emerging patterns) as interchangeable. In general, emerging pattern mining is commonly focused on discovering emerging trends amongst timestamped datasets, or extracting patterns that denote a clear and useful difference between two classes or disjoint features (male vs female, poisonous vs edible, cured vs not cured, etc). Additionally, emerging patterns may be extremely infrequent (very low support values) unlike contrast sets, which usually include a minimum support threshold value (patterns below that threshold value are

© Springer Nature Switzerland AG 2018

S. Ventura, J. M. Luna, *Supervised Descriptive Pattern Mining*,

https://doi.org/10.1007/978-3-319-98140-6_3

discarded). Finally, another major difference between contraset sets and emerging patterns is related to the metric used to quantify their interest. As described in Chap. 2, a contrast set (denoted as a pattern P) is quantified as the difference in the frequencies (support values) of P on different datasets to be contrasted. On the contrary, the interest of an emerging pattern is quantified as the ratio (also known as growth rate) of their supports in a dataset over that in the other dataset. As a result, a very promising contrast set according to the difference in support values may not be a useful emerging patterns or vice versa.

In order to understand the emerging pattern concept, let us consider the same example provided by *Dong et al.* [6] in 1999, that is, using the Mushroom dataset and a growth rate threshold of 2.5. In this example, authors described two emerging patterns ($P_1 = \{Odor = none, Gill\_Size = broad, Ring\_Number = one\}$ and $P_2 = \{Bruises = no, Gill\_Spacing = close, Veil\_Color = white\}$) described on two classes (poisonous and edible). Here, P_1 describes an odor free mushroom with a broad gill size and with only one ring; whereas P_2 denotes a non-bruising mushroom with close gill spacing and having white veil. For these two types of mushrooms, it is obtained that P_1 does not appear in the poisonous dataset, and it appears in 63.9% of the records in the edible dataset. On the contrary, P_2 is really frequent in the poisonous dataset (81.4% of the records) and very infrequent (3.8% of the records) in the edible dataset. As a result, P_1 is an emerging pattern whose growth rate is ∞, whereas the growth rate of P_2 is 21.4. What is more interesting is that even when P_1 is an emerging pattern, none of its singletons ($\{Odor = none\}$, $\{Gill\_Size = broad\}$ and $\{Ring\_Number = one\}$) are emerging patterns (their growth rates are not greater than the threshold 2.5).

From its formal definition provided by *Dong et al.* [6], many different algorithms have been proposed by various researchers. The problem of mining emerging patterns is not a trivial task since the computational cost highly depends on the dimensionality of the dataset to be analysed and, therefore, high efficient algorithms are required. Given a dataset comprising n singletons or single items, i.e. $I = \{i_1, i_2, ..., i_n\}$, a maximum of $2^n - 1$ different feasible emerging patterns can be found, so any straightforward approach becomes extremely complex with the increasing number of singletons. In a recent taxonomy [13], emerging pattern algorithms were divided into four different groups according to the strategy followed: border-based, constraint-based, tree-based and evolutionary fuzzy system-based algorithms. Some of the most important algorithms in each of the above groups are studied in this chapter. Finally, focusing ont he application domains, emerging patterns have been mainly applied to the field of bioinformatics and, more specifically, to microarray data analysis [20]. *Li et al.* [18] focused their studies on mining emerging patterns to analyse genes related to colon cancer. Emerging patterns have been also applied to the analysis of customers' behaviour [25] to discover unexpected changes in their shopping habits.

3.2 Task Definition

Let $I = \{i_1, i_2, ..., i_n\}$ be the set of items included in two datasets Ω_1 and Ω_2, which are represented in a binary tabular form (each single item may or may not appear in data). Let $T = \{t_1^1, t_2^1, ..., t_m^1\}$ be the collection of transactions or data records such as $\forall t^1 \in T : t^1 \subseteq I \in \Omega_1$. An Emerging pattern [24] is an itemset $P \subseteq I$ satisfying that the frequency of occurrence of P in Ω_1, denoted as $support(P, \Omega)$, is highly different to the frequency of the same pattern P in dataset Ω_2. This difference in support is quantified by the growth rate, that is, the ratio of the two supports ($support(P, \Omega_1)/support(P, \Omega_2)$ or $support(P, \Omega_2)/support(P, \Omega_1)$). Growth rate values greater than 1 denotes an emerging pattern, whereas values equal to 1 show uniformity in the two datasets ($support(P, \Omega_1) = support(P, \Omega_2)$). In those cases where $support(P, \Omega_1) = 0$ and $support(P, \Omega_2) \neq 0$, or $support(P, \Omega_1) \neq 0$ and $support(P, \Omega_2) = 0$, then the growth rate is defined as ∞. Some authors [24] denoted emerging patterns as association rules with an itemset (the pattern P) in the antecedent and a fixed consequent (the dataset), that is, $P \rightarrow \Omega$. Hence, the growth rate is calculated as $confidence(P \rightarrow \Omega_1)/confidence(P \rightarrow \Omega_2) = confidence(P \rightarrow \Omega_1)/(1 - confidence(P \rightarrow \Omega_1))$.

When dealing with the emerging pattern mining problem, it is important to remark that the downward closure property is not satisfied (subsets of frequent patterns are also frequent). Thus, given an emerging pattern $P \subseteq I \in \Omega$, it is possible that any of its sub-patterns $P' \subset P$ are not emerging patterns. It is therefore a non-convex problem [29] in which more general patterns may have lower growth rate values than more specific patterns. As a matter of clarification, a pattern P_1 having $support(P_1, \Omega_1) = 0.1$ and $support(P_1, \Omega_2) = 0.01$ is better than another pattern P_2 having $support(P_2, \Omega_1) = 0.7$ and $support(P_2, \Omega_2) = 0.1$ since $0.1/0.01 > 0.7/0.1$.

Finally, once the definition of emerging patterns has been given, it is important to introduce a different measure used in emerging patterns. This measure, called the strength [7], aims at reflecting the discriminative power of an emerging pattern P in Ω_i. In other words, given an emerging pattern P and a dataset Ω_i, the strength of P is mathematically defined as $strength(P, \Omega_i) = (GR(P, \Omega_i)/(GR(P, \Omega_i) + 1)) \times support(P, \Omega_i)$.

3.2.1 Problem Decomposition

According to *Dong et al.* [6], the emerging pattern mining problem can be divided into three sub-problems. To understand them, let us consider the support plane illustrated in Fig. 3.1, where x-axis represents $support(P, \Omega_1)$ for a pattern P and y-axis denotes $support(P, \Omega_2)$ for the same pattern. Here, the growth rate is considered as the variation in the support from Ω_1 to Ω_2, calculated as $GR(P) = support(P, \Omega_2)/support(P, \Omega_1) \geq \rho$, ρ being a minimum growth rate value.

From its definition, it is obtained that the growth rate takes the value 1 when $support(P, \Omega_1) = support(P, \Omega_2)$, and it is generally considered as useless since it denotes uniformity between Ω_1 and Ω_2. Based on the aforementioned definitions, any emerging pattern should be placed within the shared area (triangle ACE, see Fig. 3.1) which upper bound is given by line $y = x/\rho$ or $support(P, \Omega_1) = support(P, \Omega_2)/\rho$. At this point, it is important to highlight that those items that occur in one dataset but not the other are removed from the transactions. The decomposition of the emerging pattern mining problem is therefore described as follows:

- **BCDF rectangle**. Emerging patterns in the BCDF rectangle are those patterns whose support in Ω_1 is less than a fixed threshold δ and Ω_2 is greater or equal to a threshold θ. Mathematically it can be expressed as any pattern P satisfying $support(P, \Omega_1) < \delta$ and $support(P, \Omega_2) \geq \theta$. In this subproblem, the aim is not to use these values but the collection of all itemsets or patterns satisfying these values (those included in the borders or boundaries of the rectangle). These sets or *borders* can be obtained by means of a border-discovery algorithm such as Max-Miner [3]. Here, taking the left-hand bound L and the right-hand bound R, the collection of patterns (or itemsets) represented by the border $\langle L, R \rangle$ is $Y : \exists X \in L, \exists Z \in R$ such that $X \subseteq Y \subseteq Z$. Given $L = \{i_1, i_2 i_3\}$ and $R = \{i_1 i_2 i_3, i_2 i_3 i_4\}$, the collection of itemsets represented by the border $\langle L, R \rangle$ is $\{i_1\}, \{i_1 i_2\}, \{i_1 i_3\}, \{i_1 i_2 i_3\}, \{i_2 i_3\}, \{i_2 i_3 i_4\}$. All these sets are both supersets of either $\{i_1\}$ or $\{i_2 i_3\}$, and subsets of either $\{i_1 i_2 i_3\}$ or $\{i_2 i_3 i_4\}$.

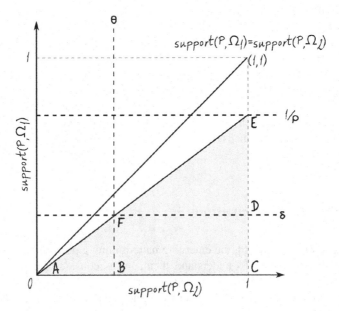

Fig. 3.1 Support plane. x-axis represents $support(P, \Omega_1)$ for a pattern P, whereas y-axis denotes $support(P, \Omega_2)$

- **DEF triangle**. This area includes all patterns whose supports in Ω_1 are greater or equal to δ and they are greater or equal to θ in Ω_2. Mathematically it can be expressed as any pattern P satisfying $support(P, \Omega_1) \geq \delta$ and $support(P, \Omega_2) \geq \theta$. The resulting set of candidates is exactly the intersection between the two borders according to $delta$ and θ. In situations where this resulting set is relative small, then emerging patterns can be found by checking the support values of all candidates in the intersection. When the set obtained from the intersection is too large, it is possible to solve this subproblem by recursively applying the same decomposition used in the triangle ACE, but considering now the triangle DEF.
- **ABF triangle**. This area includes patterns whose supports are very low in Ω_1 and Ω_2 or both. Thus, any pattern P in this area satisfies $support(P, \Omega_1) < \delta$ (or $support(P, \Omega_2) < \theta$). From the three sub-problems, it is the most challenging one and different approaches have been proposed to perform this sub-problem [6].

3.2.2 Types of Emerging Patterns

The problem of mining emerging patterns is computationally harder with respect to the number of items. Given a dataset comprising $I = \{i_1, i_2, ..., i_n\}$ items, a maximum of $2^n - 1$ different feasible emerging patterns can be found. To be able to tackle this problem in an efficient way, different authors [5] have attempted to reduce the number of extracted patterns—taking only those patterns which describe some specific characteristics. The types of emerging patterns most used in the literature are summarized below [13]:

- **Jumping emerging patterns**. These are emerging patterns that occur in one dataset but not the other, obtaining a growth rate equal to infinity. In a formal way, a pattern P is defined as a jumping emerging pattern if $support(P, \Omega_1) = 0$ and, therefore, $GR(P) = support(P, \Omega_2)/support(P, \Omega_1) = \infty$.
- **Minimal emerging patterns**. An emerging pattern P is defined as a minimal emerging pattern if any of its sub-patterns are not emerging patterns. In a formal way, P is a minimal emerging pattern if it is an emerging pattern, that is, $GR(P) \geq \rho$, and $\nexists S : S \subset P, GR(S) \geq \rho$, ρ being a minimum growth rate value. These emerging patterns are the most general ones since they contain a low number of items.
- **Maximal emerging patterns**. An emerging pattern P is defined as a maximal emerging pattern [28] if any of its super-patterns are not emerging patterns. In a formal way, P is a maximal emerging pattern if it is an emerging pattern, that is, $GR(P) \geq \rho$, and $\nexists S : S \supset P, GR(S) \geq \rho$, ρ being a minimum growth rate value.
- **Essential jumping emerging patterns**. These emerging patterns, also known as strong jumping emerging patterns, are a special case of jumping emerging

patterns that satisfy that none of their subpatterns are also jumping emerging patterns. In a formal way and considering a jumping emerging pattern P such as $GR(P) = \infty$, it is also a strong jumping emerging pattern if $\not\exists S : S \subset P, GR(S) = \infty$.

- **Noise-tolerant emerging patterns.** Let us consider two positive integers δ_1 and δ_2 such as $\delta_2 \gg \delta_1$. An emerging pattern P is a noise-tolerant emerging pattern [12] (also known as constrained emerging pattern) if $support(P, \Omega_1) \leq \delta_1$ and $support(P, \Omega_2) \geq \delta_2$.

- **Generalized noise-tolerant emerging patterns.** These patterns [12] are defined as extensions of noise-tolerant emerging patterns. Here, given two positive integers $\delta_1 > 0$ and $\delta_2 > 0$ such as $\delta_2 \gg \delta_1$, and two monotone functions f_1 and f_2, an itemset or pattern P is a generalized noise-tolerant emerging pattern from Ω_1 to Ω_2 if $f_2(support(P, \Omega_2))/f_1(support(P, \Omega_1)) \geq f_2(\delta_2)/f_1(\delta_1)$; $f_2(support(P, \Omega_2)) \geq f_2(\delta_2)$; and any subset of P does not satisfy the two previous conditions, that is, $\not\exists S \subset P : f_2(support(S, \Omega_2))/f_1(support(S, \Omega_1)) \geq f_2(\delta_2)/f_1(\delta_1)$, $f_2(support(S, \Omega_2)) \geq f_2(\delta_2)$.

- **Chi emerging patterns.** These patterns [11] represents a similar concept to the one provided by noise-tolerant emerging patterns with the single difference that a χ^2 test is introduced to improve the descriptive capacity of the pattern. Here, an emerging pattern P is defined as a chi emerging pattern if $support(P, \Omega_i) \geq \delta$, where δ is a minimum support threshold; $GR(P) \geq \rho$, where ρ is a minimum growth rate threshold; $\not\exists S \subset P : support(P, \Omega_i) \geq \delta, GR(P) \geq \rho, strength(S, \Omega_i) \geq strength(P, \Omega_i)$, where $strength(P, \Omega_i) = (GR(P)/(GR(P) + 1)) \times support(P, \Omega_i)$; and the size of P is 1, that is, it is a singletons. On the contrary, where the size of P is greater than one ($|P| > 1$), then it should be satisfied that $\forall Y : Y \subset X, |Y| = |X| - 1 \Rightarrow \chi(X, Y) \geq \eta$.

3.3 Algorithms for Mining Emerging Patterns

This section outlines a taxonomy of algorithms for mining emerging patterns [13]. Any existing algorithm can be classified within one of the following methods: border-based approaches, constraint-based methods, tree-based algorithms, and evolutionary fuzzy system-based proposals. Each of these four groups contains methods whose search strategies are similar.

3.3.1 Border-Based Algorithms

Border-based approaches is the first methodology for mining emerging patterns proposed in the literature [6]. In a recent overview [13], a border was defined as a pair $\langle L, R \rangle$ where L, also known as left-border, is a set of minimal emerging

---

Algorithm 3.1 Pseudo-code of the MBD-LLBorder algorithm

Require: $LargeBorder_\delta(\Omega_1), LargeBorder_\theta(\Omega_2)$
Ensure: $Borders$ ▷ set of borders $\langle L, R \rangle$ to obtain emerging patterns
1:   $EP \leftarrow \emptyset$
2:   $n \leftarrow$ number of items in $LargeBorder_\theta(\Omega_2)$
3:   **for** j from 1 to n **do**
4:      **if** $\nexists\, S_i^1 \in LargeBorder_\delta(\Omega_1), S_j^2 \in LargeBorder_\theta(\Omega_2) : S_i^1 \supset S_j^2$ **then**
5:         $C \leftarrow$ set of all maximal itemsets in $\{S_1^1 \cap S_j^2, S_2^1 \cap S_j^2, ..., S_m^1 \cap S_j^2\}$
6:         $Borders \leftarrow BorderDiff(\langle\{\emptyset\}, S_j^2\rangle, \langle\{\emptyset\}, C\rangle)$
7:      **end if**
8:   **end for**
9:   **return** $Borders$

Algorithm 3.2 Pseudo-code of BorderDiff procedure

Require: $\langle\{\emptyset\}, \{U\}\rangle, \langle\{\emptyset\}, \{S_1, S_2, ..., S_k\}\rangle$
Ensure: $\langle L, R \rangle$ ▷ border $\langle L, R \rangle$ to obtain emerging patterns
1:   $L \leftarrow \{x\} : x \in U - S_1\}$
2:   **for** i from 2 to k **do**
3:      $L \leftarrow \{X \cup \{x\} : X \in L, x \in U - S_i\}$
4:      remove non-minimal itemsets from L
5:      $R \leftarrow U$
6:   **end for**
7:   **return** $\langle L, R \rangle$

---

patterns, and R, also called right-border, is a set of maximal emerging patterns. Borders can be obtained by means of border-discovery algorithms, being the most well-known the Max-Miner algorithm [3]. A border is a valid pair $\langle L, R \rangle$ if each element of L and R is an antichain collection of sets, and each element of L is a subset of some element in R and each element of R is a superset of some element in L. A collection S of sets is an antichain if any set X and Y within S are incomparable sets, that is, $\forall X, Y \in S : X \nsubseteq Y, Y \nsubseteq x$. Finally, it is important to note that borders represent convex spaces so any pattern between L and R, that is, $\forall S : X, \subseteq S \subseteq Y, X \in L, Y \in R$, is an emerging pattern.

MBD-LLBorder [6] is one of the first algorithms proposed for mining emerging patterns (pseudo-code is shown in Algorithm 3.1). This algorithm is able to discover all existing emerging patterns in the BCDF rectangle shown in Fig. 3.1 by manipulating only the input borders. MBD-LLBorder works by performing a BorderDiff procedure (see Algorithm 3.2) multiple times, which obtains any emerging pattern within the aforementioned BCDF rectangle. Each BorderDiff running will use one itemset from the large border of Ω_2 (including n itemsets) and the whole large border of Ω_1 (including m itemsets) as the two input argument values. These large borders were obtained for some θ and δ, respectively, satisfying that $\theta = \rho \times \delta$.

As a matter of example, let us consider $\{i_2 i_3 i_5, i_3 i_4 i_6 i_7 i_8, i_2 i_4 i_5 i_8 i_9\}$ be the large border of Ω_1, and $\{i_1 i_2 i_3 i_4, i_6 i_7 i_8\}$ the large border of Ω_2. The algorithm will iterate two times since the number of items in the large border of Ω_2 is 2 and,

---

Algorithm 3.3 Pseudo-code of the JEPProducer algorithm

Require: $LargeBorder_{\delta}(\Omega_1), LargeBorder_{\theta}(\Omega_2)$
Ensure: $\langle L, R \rangle$
1: $L \leftarrow \emptyset$
2: $R \leftarrow \emptyset$
3: $n \leftarrow$ number of items in $LargeBorder_{\theta}(\Omega_2)$
4: **for** j from 1 to n **do**
5:     **if** $\nexists S_i^1 \in LargeBorder_{\delta}(\Omega_1), S_j^2 \in LargeBorder_{\theta}(\Omega_2) : S_i^1 \supset S_j^2$ **then**
6:         $Border \leftarrow BorderDiff(\langle\{\emptyset\}, S_j^2\rangle, \langle\{\emptyset\}, C\rangle)$
7:     **end if**
8:     $L \leftarrow L \cup$ the left bound of $Border$
9:     $R \leftarrow R \cup$ the right bound of $Border$
10: **end for**
11: **return** $\langle L, R \rangle$

---

therefore, $n = 2$ (see line 2, Algorithm 3.1). In the first iteration, that is $j = 1$, the set C, which is calculated in line 5, Algorithm 3.1, is equal to $\{i_2i_3, i_3i_4, i_2i_4\}$. After that, the MBD-LLBorder algorithm runs the $BorderDiff$ procedure with the input arguments $\{i_1i_2i_3i_4\}$ and $C = \{i_2i_3, i_3i_4, i_2i_4\}$. First, L is initialized to $L = \{\{i_1\}, \{i_4\}\}$. Then $i = 2$ (see lines 2 to 6, Algorithm 3.1), L is updated to $L = \{\{i_1\}, \{i_1i_2\}, \{i_1i_4\}, \{i_2i_4\}\}$ since $U - S_2 = \{i_1i_2\}$. L is then reduced as $L = \{\{i_1\}, \{i_2i_4\}\}$. Finally, L is updated to $L = \{\{i_1\}, \{i_1i_3\}, \{i_1i_2i_4\}, \{i_2i_3i_4\}\}$ and finally reduced as $L = \{\{i_1\}, \{i_2i_3i_4\}\}$, returning the border $\langle L, R \rangle$, L being the set $\{\{i_1\}, \{i_2i_3i_4\}\}$ and $R = \{i_1i_2i_3i_4\}$. As for the second and final iteration of the MBD-LLBorder algorithm, that is $j = 2$, the following border is obtained $\langle\{\{i_6\}, \{i_7\}, \{i_8\}\}, \{i_6i_7i_8\}\rangle$. Thus, the MBD-LLBorder algorithm (see Algorithm 3.1) returns two borders: $\langle\{\{i_1\}, \{i_2i_3i_4\}\}, \{i_1i_2i_3i_4\}\rangle$ and $\langle\{\{i_6\}, \{i_7\}, \{i_8\}\}, \{i_6i_7i_8\}\rangle$. Here, any existing emerging pattern can be derived from these borders as it was previously described in Sect. 3.2.1. Taking the first border, the collection of itemsets is $\{i_1\}, \{i_1i_2\}, \{i_1i_3\}, \{i_1i_4\}, \{i_1i_2i_3\}, \{i_1i_2i_4\}, \{i_1i_2i_3i_4\}$ and $\{i_2i_3i_4\}$. All these sets are both supersets of either $\{i_1\}$ or $\{i_2i_3i_4\}$, and subsets of $\{i_1i_2i_3i_4\}$. Taking the second border, the collection of itemsets is now: $\{i_6\}, \{i_7\}, \{i_8\}, \{i_6i_7\}, \{i_6i_8\}, \{i_7i_8\}$, and $\{i_6i_7i_8\}$.

Li et al. [22] analyzed the importance of avoiding to return to initial stages of the mining process to derive new emerging patterns when data is updated or some changes occur. According to the authors, and considering that changes are small compared to the quantity of data previously used to mine the emerging patterns, many emerging patterns remain valid for new data. Based on this issue, an incremental maintenance algorithm may compute the border of the new space efficiently by making use of the previous border space [19], not requiring a mining process from scratch. As a matter of example, and considering the insertion of new instances, it is only required to perform the union of the previous border and the new border associated with the new instances. *Li et al.* proposed the JEPProducer (see pseudo-code in Algorithm 3.3), an algorithm that is quite similar to the first algorithm proposed for mining emerging patterns, that is, MBD-LLBorder [6].

DeEPs (Decision-making by Emerging Patterns) [21] is a border-based algorithm for mining emerging patterns that is based on JEPProducer [22]. In this algorithm, and also in any of the previous algorithms based on borders, the main objective is to represent by means of borders all emerging patterns contained in data without the arduous process of computing all the subsets. In these algorithms boundaries are used since they differentiate emerging patterns and non-emerging patterns. When mining a large database, the number of resulting patterns may be extremely large for an expert to identify the most important ones. Besides, manually examining a large number of patters may cause the loss of important results. Thus, two ranking methods are considered by DeEPs to sort the results: (1) Frequency-based ranking, since the larger the frequency (the occurrence of a pattern in data) of an emerging pattern is, the more important it is. (2) Length-based ranking, since the more items matched, the pattern becomes more similar to the specific instance.

3.3.2 Constraint-Based Algorithms

Current datasets tend to include plenty of items and tons of candidate patterns. In such datasets, Apriori-like [1] approaches, considering only a support constraint during the mining process, are ineffective due to the combinatorial explosion of candidate itemsets—considering k single items in data, $2^k - 1$ different candidate patterns are possible. In this regard, and taking the search space pruning as a key point, *Zhang et al.* [30] proposed the ConsEPMiner algorithm (constraint-based emerging pattern miner). ConsEPMiner includes both inherent and external constraints to prune the search space.

Focusing on external constraints, the algorithm considers some metrics (minimum values on support, growth rate and growth rate improvement) to reduce the resulting set of emerging patterns. Given an emerging pattern P, the growth rate improvement of P and denoted as $rateimp(P)$, is defined as the minimum difference between its growth rate and the growth rate of all of its subsets. This metric is mathematically defined as $rateimp(P) = min(\forall P' \subset P : GR(P) - GR(P'))$. Growth rate improvement was defined in terms of the absolute difference on growth rates and, therefore, authors [30] also defined a metric in terms of the relative difference on growth rates. Hence, given an emerging pattern P, the relative growth rate improvement of P and denoted as $rel\_rateimp(P)$, is defined as the minimum ratio of its growth rate over that of all its subsets. In a formal way, it is defined as $rel\_rateimp(P) = min(\forall P' \subset P : GR(P') > 0, GR(P)/GR(P'))$.

As for inherent constraints, which are not given by users, they allow the search space to be pruned and the computational expenses to be reduced. Three types of constraints are included: same subset support, top growth rate and same origin. Focusing on the first type of inherent constraints, a pattern (or itemset) P on a dataset Ω satisfies the same subset support constraint if and only if $\exists P' \subset P : support(P, \Omega) = support(P', \Omega)$. The second constraint, the top growth rate of a pattern P, is such a growth rate whose value is ∞, that is, P is an emerging pattern

Fig. 3.2 A complete set
enumeration tree including
three items sorted in
lexicographical order

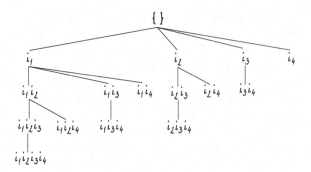

[6]. At this point, any superset of a jumping emerging pattern has P satisfies that $rateimp(P) \leq 0$ and it can therefore be pruned. Finally, itemsets containing items obtained from the same feature, and their supersets, cannot be emerging patterns and should be pruned. This constraint is named same origin constraint.

The pruning with constraints can be explained on the ordered tree shown in Fig. 3.2. In this tree, a node g is represented as a group comprising two itemsets, that is *head* or $h(g)$, which is the itemset ordered at g; and *tail* or $t(g)$, which is an ordered set consisting of itemsets that can be appended to $h(g)$ to give rise to an itemset ordered by some sub-node of g. As a matter of clarification and taking the root node from Fig. 3.2, then $h(g) = \emptyset$ and $t(g) = \{i_1, i_2, i_3, i_4\}$. Here, an itemset g' is defined as derivable from g if $h(g) \subset g'$ and $g' \subseteq h(g) \cup t(g)$.

According to *Zhang et al.* [30], pruning with inherent constraints a tree as the one shown in Fig. 3.2 is as follows. Given a group g and an item $i \in t(g)$, then i can be pruned if any of the following conditions are satisfied: $support(h(g) = support(h(g) \cup \{i\})$; or $GR(h(g) \cup \{i\} = \infty$; or both $h(g)$ and the item i have the same origin. Additionally, inherent constraints are also useful to reduce the runtime required to scan the database. Given a group g and an item $i \in t(g)$ then its child group g_c such as $h(g_c) = h(g) \cup \{i\}$ and $t(g_c) = \{j \in t(g) : j \ follows \ i\}$ satisfies that: if $support(h(g)) = support(h(g) \cup \{i\})$ then it is also satisfied that $support(h(g_c) \cup \{j\}) = support(h(g) \cup \{j\})$; and if $support(h(g)) = support(h(g) \cup \{j\})$ then it is also satisfied that $support(h(g_c) \cup \{j\}) = support(h(g) \cup \{i\})$.

Finally, considering external constraints, and taking a previously known group g satisfying that $support(h(g)) > 0$ and $support(h(g) \cup t(g))$ is not available, then the lower bound of $support(h(g) \cup t(g))$ can be derived as follows:

- $support(h(g) \cup t(g)) = 0$ in situations where $\exists G \subseteq t(g) : support(G) = 0$.
- Given a parent g_p of g, then $support(h(g_p) \cup t(g_p)) \leq support(h(g) \cup t(g))$.
- Given a parent g_p of g, then $support(h(g)) - SumDiff \leq support(h(g) \cup t(g))$, where $SumDiff$ is defined as $\Sigma_{i \in t(g)}(support(h(g_p)) - support(h(g_p) \cup \{i\}))$.

One of the most well-known algorithms for mining emerging patterns based on constraints is ConsEPMiner (see Algorithm 3.4), which is based on the Dense-

---

Algorithm 3.4 Pseudo-code of the ConsEPMiner algorithm

Require: $\Omega_1, \Omega_2, min_s, min_{GR}, min_{rateimp}$ ▷ Datasets and minimum values for support,
     growth rate and growth rate improvement
Ensure: EP ▷ set of discovered emerging patterns
1: $EP \leftarrow \emptyset$
2: $G \leftarrow$ generate initial groups from Ω_1 and Ω_2
3: **while** $G \neq \emptyset$ **do**
4:      scan Ω_2 to process the groups in G
5:      prune(G) including groups processed in Ω_2
6:      scan Ω_1 to process the groups in G
7:      **for all** $g \in G, i \in t(g)$ **do**
8:          $EP \leftarrow EP \{h(g) \cup \{i\} : h(g) \cup \{i\}$ satisfies $(min_s, min_{GR}, min_{rateimp})\}$
9:      **end for**
10:      prune(G) including groups processed in Ω_1 and Ω_2
11:      $G \leftarrow$ generate groups by expanding from G
12:      $G \leftarrow$ prune(G)
13: **end while**
14: **return** EP

Algorithm 3.5 Pseudo-code of the Prune procedure

Require: G ▷ Group to be pruned
1: **for all** $g \in G$ **do**
2:      **for all** $i \in t(g)$ **do**
3:          **if** $sameOrigin(h(g), i)$ or $support(h(g) \cup \{i\}, \Omega_2) < min_s$ or $support(h(g), \Omega_1) =$
         $support(h(g) \cup \{i\}, \Omega_1)$ or $GR(h(g) \cup \{i\}) = \infty$ or $\exists e \in E : e \subset h(g) \cup \{i\}, GR(e) = \infty$
         then
4:               $t(g) \leftarrow t(g) \setminus i$ ▷ Remove i from $t(g)$
5:          **end if**
6:      **end for**
7:      Groups with bounds of growth rate improvement > 0 can be further pruned
8: **end for**

Miner [4] algorithm proposed for mining frequent patterns. ConsEPMiner works in an iterative fashion by generating new groups from existing groups in G (see lines 3 to 13, Algorithm 3.4), where those emerging patterns of G that satisfy the given thresholds (min_s for the support; min_{GR} for the growth rate; and $min_{rateimp}$ for the growth rate improvement) are stored in the set EP. Groups are iteratively generated and pruned according a prune procedure (see Algorithm 3.5) that works depending on whether the groups are evaluated on Ω_2 or either Ω_1 and Ω_2. The ConsEPMiner algorithm finishes when no new groups are generated, returning the set EP (line 14, Algorithm 3.4). According to authors [30], the superior performance of ConsEPMiner derives from the use of both external and inherent constraints during the mining stage. Based on different studies, authors asserted that ConsEPMiner is some orders of magnitude faster than Apriori-like approaches, which only consider a prunning based on support.

3.3.3   Tree-Based Algorithms

Tree structures are demonstrated to be high efficient methods for mining patterns since they reduce the database scans by considering a compressed representation and, therefore, allowing more data to be stored in main memory. This structure has been widely used in frequent pattern mining [16] as well as in contrast set mining [5], as previously described in Chap. 2. A tree representation is therefore a compressed representation via sharing with common prefixes, where each node of the tree is a singleton including the measure associated to it.

Unlike the problem of mining frequent patterns [27], the discovery of emerging patterns requires two datasets to be considered and, therefore, the tree structure needs multiple counts to be kept in each node (one per dataset). The use of tree structures for mining emerging patterns provides two major advantages: (1) transactions sharing an itemset can be merged into single nodes with counts, allowing data to be kept in memory and providing a faster access; (2) different groups of transactions can be considered in such a way that the efficiency of the mining process highly depends on how this grouping is done. In general, trees constructed for mining emerging patterns consider a sorting of the items by the growth rate measure. For a better understanding, let us consider two sample market basket datasets Ω_1 and Ω_2 (see Table 3.1). The growth rate for each of the items (*Bread*, *Beer*, *Butter* and *Diaper*) is calculated as follows: $GR(Bread) = 6/8 = 0.75$; $GR(Beer) = 2/6 = 0.33$; $GR(Butter) = 7/3 = 2.33$; and $GR(Diaper) = 5/4 = 1.25$. The list of items is therefore sorted as $Butter \prec Diaper \prec Bread \prec Beer$, and the resulting tree is formed as shown in Fig. 3.3.

One of the first algorithms based on tree-structures for mining emerging patterns was proposed by *Bailey et al.* [2]. This algorithm was defined as an hybrid method between border-based and tree-based approaches since it uses the BorderDiff procedure—it was previously described in Algorithm 3.2—to obtain jumping emerging patterns. During a depth-first traversal of the tree, this algorithm looks for nodes which contain a non-zero counter for one dataset, and zero counters for

Table 3.1 Sample market basket datasets (Ω_1 left, Ω_2 right)

Items (dataset Ω_1)	Items (dataset Ω_2)
{Bread}	{Butter}
{Bread, Beer}	{Butter, Diaper}
{Bread, Beer, Diaper}	{Bread, Beer}
{Bread, Butter}	{Bread, Butter, Diaper}
{Bread, Beer, Butter, Diaper}	{Bread, Butter, Diaper}
{Beer}	{Bread}
{Beer, Diaper}	{Bread, Beer}
{Bread, Butter}	{Butter, Diaper}
{Bread, Beer, Diaper}	{Bread, Butter, Diaper}
{Bread}	{Butter}

Fig. 3.3 Tree ordering for
datasets Ω_1 and Ω_2.
Resulting ordering is
Butter $\prec$ *Diaper* $\prec$
Bread $\prec$ *Beer*

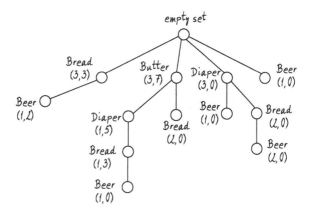

the other. Once it occurs, the itemset represented in this node is therefore a potential
jumping emerging pattern and, hence, any subset of this itemset is also potentially
a jumping emerging pattern. In this regard, a BorderDiff Procedure (it was properly
described in Sect. 3.3.1) is then run to identify all actual jumping emerging patterns
contained within the potential jumping emerging pattern.

Fan et al. [9] proposed an algorithm based on a pattern tree structure for mining
the complete set of strong jumping emerging patterns (also known as essential
jumping emerging patterns), which are those jumping emerging patterns that do
not have a subset that is also a jumping emerging pattern. Similarly to most of
the tree-based techniques for mining emerging patterns, this algorithm builds the
tree by sorting the items according to the growth rate metric. A pruning strategy is
performed in this approach, removing those subtrees that do not satisfy a minimum
support threshold. Only the most important solutions are kept by this algorithm,
which uses a filtering strategy similar to the token competition procedure. Only
those strong jumping emerging patterns that satisfy transactions that are not already
covered by better-ranked patterns are kept.

iEP-Miner [10] is another interesting algorithm that uses a tree structure to
store the dataset. This algorithm was proposed in 2003 with the aim of mining
chi emerging patterns—this type of emerging patterns was previously described in
Sect. 3.2.2. In iEP-Miner (see Algorithm 3.6) each single item in data is analyzed
and, therefore, the procedure mineSubtree (see Algorithm 3.7) is called recursively.

In 2006, *Fan et al.* [12] proposed an algorithm for mining strong jumping
emerging patterns whose support is greater than a given mininum threshold value.
For the sake of mining this type of patterns, the algorithm uses a tree structure
where the itemsets are previously ordered by the support ratio in descending order.
Given two datasets Ω_1 and Ω_2 and a minimum support threshold value min_s,
the support ratio of a pattern P was defined by the authors as $supportRatio(P)$,
taking the following values: $supportRatio(P) = 0$ if $support(P, \Omega_1) < min_s$ or
$support(P, \Omega_2) < min_s$; $supportRatio(P) = \infty$ if $support(P, \Omega_1) = 0$ and
$support(P, \Omega_2) > 0$, or if $support(P, \Omega_1) > 0$ and $support(P, \Omega_2) = 0$;
if none of the previous conditions are satisfied, then $supportRatio(P) =$
$max(support(P, \Omega_1)/support(P, \Omega_2),$ $support(P, \Omega_2)/support(P, \Omega_1))$.

Algorithm 3.6 Pseudo-code of the iEP-Miner algorithm

Require: $root, \Omega, min_s, \rho, \eta$ ▷ root of the tree structure to be traversed, dataset Ω and mininum
    thresholds for support, growth rate and chi-square
Ensure: EP ▷ set of discovered emerging patterns
  1: $EP \leftarrow \emptyset$
  2: **for all** $i \in \Omega$ **do**   ▷ analyze each single item in data
  3:     **if** $support(i, \Omega) \geq min_s$ and $GR(P) \geq \rho$ **then**
  4:         $EP \leftarrow i$
  5:         mineSubtree(i)
  6:     **end if**
  7: **end for**
  8: prune uninteresting patterns from EP
  9: **return** EP

Algorithm 3.7 Pseudo-code of the mineSubtree procedure

Require: P ▷ current pattern P
  1: **for all** $\forall j \in \Omega : j \notin P$ **do**
  2:     $P' \leftarrow P \cup j$
  3:     **if** $support(P', \Omega) \geq min_s$ and $GR(P') \geq \rho$ **then**
  4:         $EP \leftarrow P'$
  5:     **end if**
  6:     **if** $\chi(P', P) \geq \eta$ **then**
  7:         mineSubtree(P')
  8:     **end if**
  9: **end for**

At this point, let us take the aforementioned sample market basket dataset illustrated in Table 3.1 and the support ratios calculated for each of the singletons (*Bread, Beer, Butter* and *Diaper*) as follows: $supportRatio(Bread) = 1.33$; $supportRatio(Beer) = 3.00$; $supportRatio(Diaper) = 1.25$; and $supportRatio(Butter) = 2.33$. The list of items is therefore sorted as *Beer* $\prec$ *Butter* $\prec$ *Bread* $\prec$ *Diaper*, and the resulting tree is formed as shown in Fig. 3.4. Taking this resulting tree, the algorithm traverses it in a depth-first manner, checking if each node contains zero-counts in one dataset and nonzero counts in the other. Finally, the algorithm also prunes those nodes having a support value lower than a minimum threshold for the two datasets.

Terlecki et al. [26] proposed an improved method of the one proposed by *Fan et al.* [12]. The schema followed by these two algorithms are almost the same and the main difference lies on the prunning strategies. In a first prunning strategy, the new algorithm prunes the tree if the node is not D-discernibility minimal. In a formal way, a pattern P is defined as D-discernibility minimal if and only if $\forall P' \subset P : support(P, \Omega_2) < support(P', \Omega_2)$. Note that the support is calculated on second dataset. The second prunning strategy carried out by this algorithm is based on a support threshold value. The main difference with regard to other proposals is that this threshold grows dynamically during the mining process. Once the number of extracted patterns is equal to a predefined value k, all the elements are sorted in

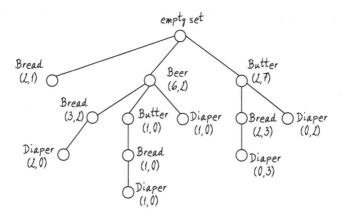

Fig. 3.4 Tree ordering for datasets Ω_1 and Ω_2. Resulting ordering is *Beer* $\prec$ *Butter* $\prec$ *Bread* $\prec$ *Diaper*

nonascending support order, and the support threshold is raised to the support of the first element plus one (absolute support values are considered).

A novel data structure, known as Dynamically Growing Contrast Pattern Tree, has been recently used by *Liu et al.* [23] for mining emerging patterns. A major feature of this structure is its ability to store a bit string representing the transactions covered by a pattern. The process of mining emerging patterns is performed in two stages: (1) singletons are taken and sorted according to the growth rate; (2) during the construction of the tree, those nodes covering new transactions and whose support values are greater than a mininum threshold are taken.

3.3.4 Evolutionary Fuzzy System-Based Algorithms

In 2017, *García-Vico et al.* [14] proposed an algorithm, known as EvAEP (Evolutionary Algorithm for Extracting Emerging Patterns), that is based on both evolutionary computation and fuzzy logic. It was therefore proposed as an evolutionary fuzzy system, which is a well-known hybridisation between fuzzy systems [17] and evolutionary computation [8]. The main advantage of this kind of methods, regarding to more accurate mathematical methods, is that the search strategy performs an efficient global search through the whole space of solutions. This search is mainly based on the principles of natural evolution and, more specifically, in the fact that organisms that are capable of acquiring resources will tend to have descendants in the future.

EvAEP was designed with the aim of extracting emerging patterns in the form of rules to describe information with respect to an interest property (target variable or class) for the experts. EvAEP employs an evolutionary algorithm where each individual (solution to the problem) represents a conjunction of pairs variable-value

---

Algorithm 3.8 Pseudo-code of the EvAEP algorithm

Require: $\Omega, maxGenerations, pSize$ ▷ Dataset, number of generations and population size
Ensure: EP ▷ set of discovered emerging patterns
 1: $EP \leftarrow \emptyset$
 2: **for all** $\Omega_i \in \Omega$ **do**   ▷ analyze each subdataset in Ω
 3:     **repeat**
 4:         $nGenerations \leftarrow 0$
 5:         $P_{nGenerations} \leftarrow$ Generate a random set of $pSize$ solutions
 6:         Evaluate solutions from $P_{nGenerations}$ considering Ω_i
 7:         **for all** $nGenerations \leq maxGenerations$ **do** ▷ Iterates for a number of generations
 8:             $P_{nGenerations+1} \leftarrow$ Take the best solution from $P_{nGenerations}$
 9:             $P_{nGenerations+1} \leftarrow P_{nGenerations+1} \cup$ Apply genetic operators in $P_{nGenerations}$
10:             Evaluate solutions from $P_{nGenerations+1}$ considering Ω_i
11:             $nGenerations \leftarrow nGenerations + 1$
12:         **end for**
13:         $S \leftarrow$ Take the best solution from $P_{nGenerations}$
14:         $EP \leftarrow EP \cup S$
15:         Mark examples in Ω_i covered by S
16:     **until** $GR \leq 1$ or no new examples are covered by S
17: **end for**
18: **return** EP

---

defined in a discrete domain. In such situations where the variable is defined in a continuous domain, EvAEP employs a fuzzy representation with fuzzy sets composed by linguistic labels defined with uniform triangles forms. In order to find high quality solutions, the set of individuals is evolved by considering an iterative rule learning model [8] that is executed once for each value of the target variable and it is repeated until a stopping criterion is reached. During each of these iterative processes, best solutions (according to a fitness function based on different quality measures) found are returned to a general set including the whole set of discovered emerging patterns.

The pseudocode of the EvAEP model is shown in Algorithm 3.8, where it is illustrated how the mining process is run for each subdataset (one per class value). For each of these iterations (see lines 2 to 17, Algorithm 3.8), EvAEP introduces an additional iterative process (see lines 3 to 16) to take the best solution found by an evolutionary process (see lines 7 to 12). Before the evolutionary process starts, EvAEP generates an initial population with a size determined through the external parameter $pSize$. In each iteration of the evolutionary process, the best solution (according to the fitness function value) is taken to be passed to the next generation, working as the elite. Next generation is also filled with new individuals obtained through the application of genetic operators (see line 9). At the end of the evolutionary process (once a number of generations is reached), the best found solution is kept (see lines 13 and 14) and the process is repeated once again (till the best found solution has a growth rate lower or equal to 1, or no new examples are covered).

Recently, the previously described EvAEP algorithm was redesigned towards distributed paradigms such as MapReduce to work in Big Data environments [15].

This new approach, called EvAEP-Spark, is development in Apache Spark to enable huge datasets to be efficiently analysed. The main idea of EvAEP-Spark is to modify the methodology carried out to evaluate individuals along the evolutionary process of the original EvAEP algorithm. In this regard, the MapReduce framework is used (in each generation of the evolutionary process) to reduce the complexity when evaluating the population or set of individuals. During the map phase, the whole dataset is divided into chunks of data, and each of these subsets is analysed in a different mapper for all the individuals in population. After that, the reduce phase is responsible for calculating the fitness function for each individual by considering the partial results obtained by the map phase.

References

1. R. Agrawal, T. Imielinski, A.N. Swami, Mining association rules between sets of items in large databases, in *Proceedings of the 1993 ACM SIGMOD International Conference on Management of Data*, SIGMOD Conference '93, pp. 207–216, Washington, DC (1993)
2. J. Bailey, T. Manoukian, K. Ramamohanarao, Fast algorithms for mining emerging patterns, in *Proceedings of the 6th European Conference on Principles of Data Mining and Knowledge Discovery*, PKDD '02, pp. 39–50, London (2002)
3. R.J. Bayardo, Efficiently mining long patterns from databases, in *Proceedings of the 1998 ACM SIGMOD International Conference on Management of Data*, SIGMOD '98, pp. 85–93, Seattle, Washington (1998)
4. R.J. Bayardo, R. Agrawal, D. Gunopulos, Constraint-based rule mining in large, dense databases. Data Min. Knowl. Disc. **4**, 217–240 (1999)
5. G. Dong, J. Bailey (eds.), *Contrast Data Mining: Concepts, Algorithms, and Applications* (CRC Press, Boca Raton, 2013)
6. G. Dong, J. Li, Efficient mining of emerging patterns: discovering trends and differences, in *Proceedings of the 5th ACM SIGKDD International Conference on Knowledge Discovery and Data Mining (KDD-99)*, pp. 43–52, New York, (1999)
7. G. Dong, X. Zhang, L. Wong, J. Li, Caep: classification by aggregating emerging patterns, in *Proceedings of the 2nd International Conference on Discovery Science (DS-99)*, pp. 30–42, Tokyo, December 1999
8. A.E. Eiben, J.E. Smith, *Introduction to Evolutionary Computing* (Springer, Berlin, 2003)
9. H. Fan, K. Ramamohanarao, A Bayesian approach to use emerging patterns for classification, in *Proceedings of the 14th Australasian Database Conference*, ADC '03, pp. 39–48, Adelaide (2003)
10. H. Fan, K. Ramamohanarao, Efficiently mining interesting emerging patterns, in *Proceedings of the 4th International Conference on Web-Age Information Management (WAIM-03)*, pp. 189–201, Chengdu, August 2003
11. H. Fan, K. Ramamohanarao, Noise tolerant classification by chi emerging patterns, in *Proceedings of the 8th Pacific-Asia Conference on Advances in Knowledge Discovery and Data Mining*, PAKDD 2004, pp. 201–206, Sydney, May 2004
12. H. Fan, K. Ramamohanarao, Fast discovery and the generalization of strong jumping emerging patterns for building compact and accurate classifiers. IEEE Trans. Knowl. Data Eng. **18**(6), 721–737 (2006)
13. A.M. García-Vico, C.J. Carmona, D. Martín, M. García-Borroto, M.J. del Jesus, An overview of emerging pattern mining in supervised descriptive rule discovery: taxonomy, empirical study, trends and prospects. Wiley Interdiscip. Rev. Data Min. Knowl. Disc. **8**(1) (2018)

14. A.M. García-Vico, J. Montes, J. Aguilera, C.J. Carmona, M.J. del Jesus, Analysing concentrating photovoltaics technology through the use of emerging pattern mining, in *Proceedings of the 11th International Conference on Soft Computing Models in Industrial and Environmental Applications (SOCO 16)*, pp. 334–344, San Sebastian, October 2016

15. A.M. García-Vico, P. González, M.J. del Jesús, C.J. Carmona, A first approach to handle fuzzy emerging patterns mining on big data problems: the EvAEFP-spark algorithm, in *Proceedings of the 2017 IEEE International Conference on Fuzzy Systems*, pp. 1–6, Naples, July 2017

16. J. Han, J. Pei, Y. Yin, R. Mao, Mining frequent patterns without candidate generation: a frequent-pattern tree approach. Data Min. Knowl. Disc. **8**, 53–87 (2004)

17. F. Herrera, Genetic fuzzy systems: taxonomy, current research trends and prospects. Evol. Intell. **1**(1), 27–46 (2008)

18. J. Li, L. Wong, Identifying good diagnostic gene groups from gene expression profiles using the concept of emerging patterns. Bioinformatics **18**(10), 1406–1407 (2002)

19. J. Li, K. Ramamohanarao, G. Dong, The space of jumping emerging patterns and its incremental maintenance algorithms, in *Proceedings of the 17th International Conference on Machine Learning (ICML 2000)*, Stanford, CA, June 2000, pp. 551–558

20. J. Li, H. Liu, J.R. Downing, A.E. Yeoh, L. Wong, Simple rules underlying gene expression profiles of more than six subtypes of acute lymphoblastic leukemia (ALL) patients. Bioinformatics **19**(1), 71–78 (2003)

21. J. Li, G. Dong, K. Ramamohanarao, L. Wong, DeEPs: a new instance-based lazy discovery and classification system. Mach. Learn. **54**(2), 99–124 (2004)

22. J. Li, T. Manoukian, G. Dong, K. Ramamohanarao, Incremental maintenance on the border of the space of emerging patterns. Data Min. Knowl. Disc. **9**(1), 89–116 (2004)

23. Q. Liu, P. Shi, Z. Hu, Y. Zhang, A novel approach of mining strong jumping emerging patterns based on BSC-tree. Int. J. Syst. Sci. **45**(3), 598–615 (2014)

24. P.K. Novak, N. Lavrač, G.I. Webb, Supervised descriptive rule discovery: a unifying survey of contrast set, emerging pattern and subgroup mining. J. Mach. Learn. Res. **10**, 377–403 (2009)

25. H.S. Song, J.K. Kimb, H.K. Soung, Mining the change of customer behavior in an internet shopping mall. Expert Syst. Appl. **21**(3), 157–168 (2001)

26. P. Terlecki, K. Walczak, Efficient discovery of top-k minimal jumping emerging patterns, in *Proceedings of the 6th International Conference on Rough Sets and Current Trends in Computing*, RSCTC 2008, pp. 438–447, Akron, October 2008

27. S. Ventura, J.M. Luna, *Pattern Mining with Evolutionary Algorithms* (Springer International Publishing, Berlin, 2016)

28. Z. Wang, H. Fan, K. Ramamohanarao, Exploiting maximal emerging patterns for classification, in *Proceedings of the 17th Australian Joint Conference on Artificial Intelligence*, pp. 1062–1068, Cairns, December 2004

29. L. Wang, H. Zhao, G. Dong, J. Li, On the complexity of finding emerging patterns. Theor. Comput. Sci. **335**(1), 15–27 (2005)

30. X. Zhang, G. Dong, K. Ramamohanarao, Exploring constraints to efficiently mine emerging patterns from large high-dimensional datasets, in *Proceedings of the 6th ACM SIGKDD International Conference on Knowledge Discovery and Data Mining, ACM SIGKDD '00*, pp. 310–314, Boston, August 2000

Chapter 4
Subgroup Discovery

Abstract Subgroup discovery is the most well-known task within the supervised descriptive pattern mining field. It aims at discovering patterns in the form of rules induced from labeled data. This chapter therefore introduces the subgroup discovery problem and also describes the main differences with regard to classification and clustering tasks. Additionally, it provides a good description about similarities and differences with respect to other well-known tasks within the supervised descriptive pattern mining field such as contrast set mining and emerging pattern mining. Finally, the most widely used metrics in this field as well as important approaches to perform this task are analysed.

4.1 Introduction

Subgroup discovery is considered as a general and broadly applicable data mining technique for descriptive and exploratory analysis. It aims at identifying some interesting relationships between patterns or variables in a set with respect to a specific target variable. The information extracted by this data mining technique is normally represented in the form of rules according to the user's interest measured by different quality functions. For these rules, conditions included in the antecedent are used to identify transactions for which a certain property in the target item (or set of items) holds. Thus, subgroup discovery is a task that lies at the intersection of predictive and descriptive induction [36] and this and other tasks (contrast set mining [7] and emerging pattern mining [21]) are grouped under the heading of *supervised descriptive pattern mining*.

In a recent overview [25] the usefulness of discovering subgroups with respect to a specific target variable was described by the following example. Let us consider a dataset comprising information about age (less than 25, between 25 and 60, and more than 60), sex (male or female) and country (Spain, USA, France and German) for a group of people. Let us also introduce a variable of interest related to the money (poor, average and rich). In such a dataset, it is feasible to discover some subgroups in the form of rules containing the following descriptions: (1) IF (Age = Less than 25 AND Country = German) THEN Money = Rich. It denotes a subset of

© Springer Nature Switzerland AG 2018

S. Ventura, J. M. Luna, *Supervised Descriptive Pattern Mining*,

https://doi.org/10.1007/978-3-319-98140-6_4

German people with less than 25 years old for which the probability of being rich is unusually high with respect to the rest of the population. (2) IF (Age = More than 60 AND Sex = Female) THEN Money = Average. It represents a set of women with more than 60 years old that are more likely to have an average economy than the rest of the population.

Subgroup discovery is therefore a supervised descriptive pattern mining task that is defined halfway between supervised and unsupervised learning tasks. The former aims at obtaining an accurate classifier that predicts the correct output given an input. On the contrary, the goal of unsupervised learning tasks is to describe hidden structures by exploring unlabeled data. When comparing the subgroup discovery task with other supervised and unsupervised learning tasks (classification and clustering), it is obtained that subgroup discovery seeks to extract interesting subgroups (not an accurate classifier as classification does) for a given class attribute and the discovered subgroups do not necessarily cover all the examples for the class. Thus, it aims at describing data subsets by means of independent and simple rules unlike predictive learning which aims at forming complex models that explain future behaviour. On the other hand, unlike clustering that aims to describe unlabeled data structures, subgroup discovery is conceived to form subsets that similarly behave, so new input data could be classified according to these subgroups. Figure 4.1 graphically illustrates the difference between the classification, clustering and subgroup discovery tasks.

Subgroup discovery was first defined by *Klosgen et al.* [28] in 1996 and since then, this task has been well investigated concerning binary and nominal target variables [25]. Additionally, numeric target variables have received increasing attention recently [22]. For any of these target variables, a wide number of approaches have been proposed, following either deterministic and stochastics methodologies. Early subgroup discovery approaches were based on deterministic models, consisting in adapting either association rule mining or classification algorithms for the subgroup discovery task [26]. Then, many authors [10, 17] have focused their studies on evolutionary algorithms to reduce the computational time and to increase the interpretability of the resulting model. A major feature of any of the evolutionary algorithms proposed to this aim is their ability to discover reliable and highly representative subgroups with a low number of rules and variables per rule. Finally, as a consequence of the Big Data era, a number of algorithms based on the parallel and distributed paradigms have been proposed [39]. These and many other algorithms in this field are properly described in this chapter.

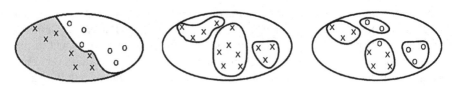

Fig. 4.1 Difference between classification (left), clustering (center) and subgroup discovery (right) tasks

4.2   Task Definition

The concept of subgroup discovery was first introduced by *Klösgen* [28] and *Wrobel* [44] as follows:

> Given a population of individuals (customers, objects, etc.) and a property of those individuals that we are interested in, the task of subgroup discovery is to find population subgroups that are statistically most interesting for the user, e.g. subgroups that are as large as possible and have the most unusual statistical characteristics with respect to a target attribute of interest.

Subgroup discovery combines features of both supervised and unsupervised learning tasks [8], and uncovers explicit subgroups via single and simple rules, that is, having a clear structure and few variables [25]. In a formal way, let us assume a dataset Ω comprising a set of items $I = \{i_1, i_2, ..., i_n\} \in \Omega$ and a target variable $t \in \Omega$. This dataset is represented as a set of records $r \in \Omega$ in the form $r = \{i_1, ..., i_k, t\} : k \in \mathbb{N}^+$, t denoting the target variable. A subgroup is represented in the form $P \rightarrow t$, where $P \subseteq I$, and the set of records included in such subset is given by $\{\forall r^i \in \Omega \ : \ P \cup t \subseteq r^i\}$. These rules ($P \rightarrow t$) denote an unusual statistical distribution of P (pattern including a set of features) with respect to the target concept t or variable of interest. Since its definition, subgroup discovery has been well investigated concerning binary and nominal target variables, i.e. the property of interest is defined within a finite number of possible values [4]. Nevertheless, target concepts on continuous domains have received increasing attention in recent years [4].

Regardless the type of target variable, either taken from a discrete or continuous domain, the choice of a good quality measure to denote the unusual statistical distribution is a key point as it has been discussed by many researchers [19]. This statistical unusualness in the distribution is the common element of many additional supervised descriptive pattern mining tasks, e.g. contrast set mining and emerging patterns, so some differences and similarities are precisely based on such quality measures. A description of these metrics as well as an extensive comparison of subgroup discovery and other related tasks are provided below.

4.2.1   Quality Measures

When evaluating a set of subgroups extracted by any subgroup discovery algorithm, the choice of a good quality measure to each particular problem is essential [19]. These quality measures can be divided into two main groups, depending on the target variable.

4.2.1.1   Discrete Target Variables

According to recent research articles [25], there is no clear consensus about which specific quality measure is more appropriate in each case and just a taxonomy of metrics has been proposed, including four main groups: measures of complexity, generality, precision and interest. There is also an additional group of measures that are hybrid metrics defined to obtain a tradeoff between generality, interest and precision.

- Measures of complexity are those related to the interpretability of the subgroups, i.e. the simplicity of the knowledge extracted from the subgroups. Two metrics are generally considered by researchers, that is, number of rules and number of variables. The number of rules or subgroups is related to the size of the resulting set of solutions given by the algorithms. The higher this size, the more difficult for the user is to be understand the provided knowledge. On the contrary, the number of variables is related to the size of P (also denoted as $|P|$), that is, the number of items (or elements) included in P. Given a resulting set of solutions $S = \{P_1, P_2, ..., P_m\}$, the average number of variables is therefore calculated as $\forall P_i \in S, \Sigma |P_i|/m$.
- Generality measures are used to quantify the quality of subgroups according to the set of records included in the subset. Generality measures are mainly represented by support, either considering the whole dataset or the subset represented by the target variable, and coverage. Regarding the support quality measure, it is considered as one of the most commonly used measures in pattern mining. This metric (see Eq. (4.1)) calculates the number of records included in the subset represented by the rule $P \rightarrow t$.

$$support(P \rightarrow t) = \frac{|\{\forall r^i \in \Omega \ : \ P \cup t \subseteq r^i\}|}{|\{\forall r^i \in \Omega\}|} \quad (4.1)$$

Finally, coverage (see Eq. (4.2)) is a generality measure that calculates the percentage of records of the dataset covered by P (not considering t) from the whole set of records in data.

$$coverage(P \rightarrow t) = \frac{|\{\forall r^i \in \Omega \ : \ P \subseteq r^i\}|}{|\{\forall r^i \in \Omega\}|} \quad (4.2)$$

- Measures of precision are those related to the accuracy of discovered subgroups, determining whether they are reliable enough or not. One of the most commonly used measures in this group of metrics is the confidence of each rule $P \rightarrow t$. This measure, also defined as accuracy in the specialised bibliography, determines the reliability by calculating the relative frequency of records that satisfy the complete rule among those satisfying only P (see Eq. (4.3)).

$$confidence(P \rightarrow t) = \frac{|\{\forall r^i \in \Omega \; : \; P \cup t \subseteq r^i\}|}{|\{\forall r^i \in \Omega \; : \; P \subseteq r^i\}|} \qquad (4.3)$$

Additionally, Q_c (see Eq. (4.4)) is a precision measure that quantifies the tradeoff between the number of records in the subset denoted by the rule $P \rightarrow t$ and those records in the subset given by P but do not include t, and considering a generalisation constant stated as c.

$$Q_c(P \rightarrow t) = |\{\forall r^i \in \Omega \; : \; P \cup t \subseteq r^i\}| - \qquad (4.4)$$

$$-c \times |\{\forall r^i \in \Omega \; : \; P \subseteq r^i, t \nsubseteq r^i\}|$$

Finally, Q_g is another precision measure that determines the tradeoff of a subgroup between the number of transactions classified perfectly and the unusualness of their distribution (see Eq. (4.5)). Note that g is used as a generalisation parameter that usually takes values between 0.5 and 100.

$$Q_g(P \rightarrow t) = \frac{|\{\forall r^i \in \Omega \; : \; P \cup t \subseteq r^i\}|}{|\{\forall r^i \in \Omega \; : \; P \subseteq r^i, t \nsubseteq r^i\}| + g} \qquad (4.5)$$

- Measures of interest are those used for selecting and ranking solutions according to their potential interest to the user. Within this type of measures can be found metrics such as significance and novelty. Significance measure indicates the significance of a finding measured by the likelihood ratio of a rule $P \rightarrow t$, and it is calculated as shown in Eq. (4.6). Here, n is the number of different target values in Ω.

$$significance(P \rightarrow t) = 2 \times \sum_{j=0}^{n} |\{\forall r^i \in \Omega \; : \; P \cup t_j \subseteq r^i\}| \times \qquad (4.6)$$

$$\times \log \frac{|\{\forall r^i \in \Omega \; : \; P \cup t_j \subseteq r^i\}|}{|\{\forall r^i \in \Omega \; : t_j \subseteq r^i\}| \times \frac{|\{\forall r^i \in \Omega \; : \; P \subseteq r^i\}|}{|\{\forall r^i \in \Omega\}|}}$$

As for the novelty, this quality metric was defined to detect unusual subgroups and it was mathematically defined as shown in Eq. (4.7).

$$novelty(P \rightarrow t) = |\{\forall r^i \in \Omega \; : \; P \cup t \subseteq r^i\}| - \qquad (4.7)$$

$$-(|\{\forall r^i \in \Omega \; : \; P \subseteq r^i\}| \times |\{\forall r^i \in \Omega \; : \; t \subseteq r^i\}|)$$

- Hybrid measures are metrics that lies somewhere between generality, interest and precision. One example of this kind of measures is the unusualness of a rule, defined as the weighted relative accuracy (WRAcc) and computed as depicted in Eq. (4.8). The unusualness of a rule $P \rightarrow t$ might be described as the balance between its coverage (see Eq. (4.2)) and its accuracy gain denoted as $\frac{|\{\forall r^i \in \Omega \,:\, P \cup t \subseteq r^i\}|}{|\{\forall r^i \in \Omega\}|} - \frac{|\{\forall r^i \in \Omega \,:\, t \subseteq r^i\}|}{|\{\forall r^i \in \Omega\}|}$. Sometimes, the unusualness of the rule $P \rightarrow t$ is required to be normalized to the interval [0, 1] as illustrated in Eq. (4.9). Here, LB_{WRAcc} is calculated as $(1 - \frac{|\{\forall r^i \in \Omega \,:\, t \subseteq r^i\}|}{|\{\forall r^i \in \Omega\}|}) * (0 - \frac{|\{\forall r^i \in \Omega \,:\, t \subseteq r^i\}|}{|\{\forall r^i \in \Omega\}|})$ and UB_{WRAcc} is calculated as $\frac{|\{\forall r^i \in \Omega \,:\, t \subseteq r^i\}|}{|\{\forall r^i \in \Omega\}|} * (1 - \frac{|\{\forall r^i \in \Omega \,:\, t \subseteq r^i\}|}{|\{\forall r^i \in \Omega\}|})$.

$$WRAcc(P \rightarrow t) = \frac{|\{\forall r^i \in \Omega \,:\, P \subseteq r^i\}|}{|\{\forall r^i \in \Omega\}|} \times \tag{4.8}$$

$$\times \left(\frac{|\{\forall r^i \in \Omega \,:\, P \cup t \subseteq r^i\}|}{|\{\forall r^i \in \Omega \,:\, P \subseteq r^i\}|} - \frac{|\{\forall r^i \in \Omega \,:\, t \subseteq r^i\}|}{|\{\forall r^i \in \Omega\}|} \right)$$

$$NWRAcc(P \rightarrow t) = \frac{WRAcc(P \rightarrow t) - LB_{WRAcc}}{UB_{WRAcc} - LB_{WRAcc}} \tag{4.9}$$

This section presents a concise and comprehensive survey on interestingness measures for numerical target concepts.

Specificity is another hybrid quality measure, which determines the proportion of negative cases incorrectly classified and it is computed as shown in Eq. (4.10).

$$specificity(P \rightarrow t) = \frac{|\{\forall r^i \in \Omega \,:\, P \cup t \not\subseteq r^i\}|}{|\{\forall r^i \in \Omega \,:\, t \not\subseteq r^i\}|} \tag{4.10}$$

Another quality measure categorized under this group of hybrid metrics is the sensitivity or support on the basis of the records of the target feature. Sensitivity combines precision and generality related to the target variable. Instead of considering the whole dataset, this measure takes into account the set of records that satisfy the target variable (see Eq. (4.11)).

$$sensitivity(P \rightarrow t) = \frac{|\{\forall r^i \in \Omega \,:\, P \cup t \subseteq r^i\}|}{|\{\forall r^i \in \Omega \,:\, t \subseteq r^i\}|} \tag{4.11}$$

4.2.1.2 Numerical Target Variables

Many interestingness measures extract certain data characteristics when numerical target concepts are analysed. These quality measures are categorized with respect to the used data characteristics as follows [30].

- Mean-based interestingness measures. These metrics score subgroups by comparing the mean value in the subgroup μ_P with the mean value in the overall dataset μ_Ω. A subgroup is then considered as interesting, if the mean of the target values is significantly higher within the subgroup. In that direction, several interestingness measures have been proposed, e.g. generic mean-based function, which is computed as $q_{mean}(P) = i_P \times (\mu_P - \mu_\Omega)$, i_P denoting the number of records covered by P. Another example is the generic symmetric mean-based function, which computes the interest either as the absolute difference $q_{sym}(P) = i_P \times |\mu_P - \mu_\Omega|$ or the squared difference $q_{sqd}(P) = i_P \times (\mu_P - \mu_\Omega)^2$. Finally, variance reduction $q_{vr}(P) = (i_P/(i_\Omega - i_P)) \times (\mu_P - \mu_\Omega)^2$ is also good function to measure the interest in numerical target variables, i_Ω denoting the number of records in Ω.

- Variance-based measures. These measures were proposed to identify subgroups with an unusual variance. Within this group of measures, the generic variance-based function is computed as $q_{sq}(P) = i_P \times (\sigma_P - \sigma_\Omega)$; σ_P and σ_Ω being the standard deviations of the subgroup and the whole dataset, respectively. Another example of quality measure is t-score, computed as $q_t(P) = (\sqrt{i_P} \times (\mu_P - \mu_\Omega))/\sigma_P$.

- Median-based measures. These metrics are based on the median of target value of subgroups since the mean value is known to be sensitive to outliers. Generic median-based measure is the most well-known and it is calculated as $q_{med}(P) = i_P \times (med_P - med_\Omega)$, where med_P is the median of target values in the subgroup and med_Ω the median in the dataset.

- Rank-based measures. A variety of statistical tests for the deviation of numerical variables use the ranks of the target variable instead of the target values themselves. In these quality measures, the highest target value is assigned to the rank number one, the second highest target value is mapped to the rank number two, and so on. A major feature of such metrics is they reduce the sensitivity to outliers compared to mean-based metrics.

- Distribution-based measures. This group includes a metric that captures increases as well as decreases of the target values, which is defined as $q_{ks}(P) = \sqrt{(i_P \times i_{\neg P})/i_\Omega} \times \Delta(P, \neg P)$, where $\Delta(P, \neg P)$ is the supremum of differences in the empirical distribution function induced by the subgroup P and its complement $\neg P$.

4.2.2 Unifying Related Tasks

Supervised descriptive pattern mining, also known as supervised descriptive rule discovery [36] due to knowledge is represented in the form of rules, gathers a group of techniques where the main proposal is the search for interesting descriptions in data with respect to a property or variable of interest. The most representative techniques within supervised descriptive pattern mining are subgroup discovery [25], contrast set mining [18] (see Chap. 2) and emerging pattern mining [21] (see Chap. 3).

When analysing the definitions of the three aforementioned techniques, that is, contrast set mining, emerging pattern mining and subgroup discovery, some differences appear and they are mainly based on the terminology and their quality measures. As a general description, contrast set mining searches for discriminating characteristics of groups called contrast sets; emerging pattern mining aims at discovering itemsets whose support increases significantly from one dataset to another; whereas subgroup discovery searches for data subset descriptions. According to authors in [36], definitions of two dissimilar tasks are compatible if one of these tasks can be translated into the other one without substantially changing the goal. At this point, and based on the previous information, it can be demonstrated that definitions of contrast set mining, emerging pattern mining and subgroup discovery tasks are compatible.

Taking the contrast set mining task restricted to consider only two groups of records, then it is equivalent to the task of mining emerging patterns where two datasets are compared. Let us consider a pattern P and two datasets (Ω_1 and Ω_2), these sets can also be seen as groups of records ($G_1 \equiv \Omega_1$ and $G_2 \equiv \Omega_2$) so the support of P in these datasets (or groups of records) can be simplified to $support(P, \Omega_1)$ and $support(P, \Omega_2)$. Hence, the main difference between emerging patterns and contrast sets lies in the quality of P that is calculated as the ratio of the two supports ($support(P, \Omega_1)/support(P, \Omega_2)$ or $support(P, \Omega_2)/support(P, \Omega_1)$) for emerging patterns, and the difference in the frequencies, that is, $|support(P, \Omega_1) - support(P, \Omega_2)|$, for contrast set mining. Additionally, comparing two-groups contrast set mining (or emerging pattern mining since it was demonstrated that they are compatible) may also be directly translated into a subgroup discovery task where the target variable includes two values. Here, each target value (t_1 and t_2) denotes therefore a different and disjoint group of records and both tasks, that is, subgroup discovery and contrast set mining, can be defined as compatible.

In a recent analysis [12], it was revealed that there is a real compatibility between subgroup discovery, emerging patterns and contrast sets thanks to the use of the weighted relative accuracy (WRAcc, see Eq. (4.8)). In this analysis, authors demonstrated that both WRAcc (for the subgroup discovery task) and GR (for the emerging pattern mining task) obtain the same value when WRAcc is greater than zero. Therefore, they concluded that all patterns with positive values of WRAcc also represent emerging patterns. Additionally, and considering the normalized weighted related accuracy (NWRAcc, see Eq. (4.9)), denoting good levels when values are greater than 0.5, authors in [12] demonstrated that NWRAcc values ≥ 0.55 also represent a contrast set; whereas NWRAcc values ≥ 0.5 denotes emerging patterns. As a summary, *Carmona et al.* [12] described arguments to show the direct relationships between subgroup discovery, emerging patterns and contrast sets through the use of the WRAcc quality measure.

4.3 Algorithms for Subgroup Discovery

Subgroup discovery is the most well-known supervised descriptive pattern mining task and numerous algorithms have been proposed in literature [25] to perform this task. Existing proposals can be categorized in groups according to a series of elements as it is described below [6]:

- Type of the targets. The target variable can be defined in different domains (binary, continuous, nominal) and the analysis is therefore required to be performed on different ways. Binary analysis represents target variables with only two values (True or False) where interesting subgroups for each of the possible values are required to be obtained. Following the same philosophy of binary analysis, when target variables are defined in a discrete domain, the aim is to find subgroups for each of the feasible values. Finally, when numeric target values are considered, a more complex study should be carried out by dividing the variable in two ranges with respect to the average, discretising the variable in a determined number of intervals, or searching for significant deviations of the mean.
- Description language. The subgroup representation is a key point since, in order to be as interesting as possible, it is required to be simple and represented in an understandable form. According to different authors [25], variables that define the subgroup can be represented in a positive/negative form, through fuzzy logic, considering ranges of values, etc.
- Quality measures. It is important to denote the right metrics that provide the expert with the importance and interest of the obtained subgroups. Different quality measures have been presented in the specialised bibliography [19], most of them were properly described in Sect. 4.2.1 of this Chapter. It is important to remark that there is no consensus about which metric (or set of metrics) should be used in subgroup discovery.
- Search strategy. This is a relevant element in subgroup discovery since the dimension of the search space has an exponential relation to the number of features and values to be analysed. The use of the right search strategy for each specific problem is essential to perform the subgroup discovery task in a desiderable time.

The most widely used subgroup discovery algorithms are described below, which are divided into four main groups: those based on extensions of classification algorithms [34]; extensions of association rule mining [2]; evolutionary algorithms [43]; and those algorithms properly designed to be run on Big Data environments.

4.3.1 Extensions of Classification Algorithms

The first algorithms developed so far for the subgroup discovery task were defined as extensions of classification algorithms, considering decision trees to represent the extracted knowledge. EXPLORA [28] was the first approach developed for the discovery of interesting subgroups. It was based on defining decision trees through a statistical verification method, so the interest of the results was measured using statistical metrics. MIDOS [44] was second proposal for mining interesting subsets, and it was defined as an extension of EXPLORA [28] to multi-relational databases.

The principal aim was to discover subgroups of a given target variable that present an unusual statistical distribution with respect to the complete population.

Since the two aforementioned approaches were proposed, many additional algorithms based on classification have been proposed for the subgroup discovery task [20, 29]. All these algorithms were designed to discover individual rules (each rule represents a subgroup) of interest instead of building accurate models from a set of rules as classification rule learning algorithms do. SubgroupMiner [27] was proposed as an extension of both EXPLORA and MIDOS algorithms. More than a specific algorithm, it is an advanced subgroup discovery system that efficiently integrates large databases and different visualisation methods. The algorithm for mining subgroups verifies whether the statistical distribution of the target (nominal target variable) is significantly different in the extracted subgroup.

Gamberger et al. [20] proposed the SD algorithm that searches for solutions that maximize Q_g (see Eq. (4.5)), a precision measure that determines the tradeoff of a subgroup between the number of transactions classified perfectly and the unusualness of their distribution. SD is a rule induction approach (see Algorithm 4.1) based on a variation of the beam search. This is an iterative algorithm that aims to keep the best subgroup descriptions in a fixed width beam. The algorithm initializes all the solutions (rules that represent subgroups) in *Beam* and *newBeam* with empty rule conditions. Their quality values are also initialized to zero (see lines 3 to 7, Algorithm 4.1). After the rule initialization process, an iterative procedure is performed that stops when it is no longer possible to improve the quality of the

---

Algorithm 4.1 Pseudo-code of the SD algorithm

Require: $\Omega, L, mSupport, bWidth$ ▷ dataset Ω, set L of all feasible attribute values, minimum
    support to accept a subgroup, and maximum number of rules in the beam
Ensure: S ▷ set S of the best solutions
  1: $Beam \leftarrow \emptyset$
  2: $newBeam \leftarrow \emptyset$
  3: **for** i from 1 to $bWidth$ **do**
  4:     $Cond(i) \leftarrow \emptyset$ ▷ initialize condition part of the rule to be empty
  5:     $Q_g(i) \leftarrow 0$ ▷ initialize the quality measure
  6:     $Beam \leftarrow Cond(i), Q_g(i)$
  7: **end for**
  8: **while** there are improvements in *Beam* **do**
  9:     **for** $\forall b \in Beam$ **do**   ▷ for all rules in *Beam*
10:         **for** $\forall l \in L$ **do**   ▷ for all the feasible attribute values
11:             $b \leftarrow b \wedge l$ ▷ add a new feature to the solution b in *Beam*
12:             Compute the quality measure $Q_g(b)$ for the new solution on Ω
13:             **if** $support(b) \geq mSupport \wedge Q_g(b)$ is greater than any $b \in newBeam$ **then**
14:                 The worst rule in *newBeam* is replaced by b
15:             **end if**
16:         **end for**
17:     **end for**
18:     $Beam \leftarrow newBeam$
19: **end while**
20: **return** $\langle L, R \rangle$

---

solutions in *Beam*. Rules can be improved only by conjunctively adding features from the whole set L of feasible conditions (attribute value pairs). For every new rule, constructed by conjunctively adding a feature to the existing rule (see line 11) the quality measure Q_g is computed (see line 12). If the support of the new solution is greater or equal to a minimum predefined support value, and its quality measure Q_g has been improved (it is greater than the one of any rule in *newBeam*), then the solution is considered as promising and it should be kept in the set *newBeam* (see lines 13 to 15, Algorithm 4.1). When the algorithm terminates, the first rule in *Beam* is the rule with maximum Q_g.

CN2-SD [29] is another example of an approach based on classification algorithms. It was proposed an adaptation of the classic CN2 classification algorithm [13], which is in an iterative algorithm that searches for a conjunction of attributes that behaves well according to the information-theoretic entropy measure. Thus, rules covering a large number of data records for a single target value are preferable to those covering few records of other target value. The CN2 algorithm works in an iterative fashion by keeping rules that covers transactions that were not previously covered, and this iterative process is carried out till no more satisfactory rules are found. Since CN2-SD is a modified version of CN2, some additional features should be considered: (1) a new heuristic that combines generality (measured as the support of the rule) and accuracy (quantified as the reliability of the rule) is considered; (2) weights are incorporated to the covering step so data records that were already covered are weighted in a decreasing manner; (3) a probabilistic classification based on the target variable distribution of the covered data records is considered. Finally, it is important to remark that a major drawback of CN2-SD is that it only works with binary target variables (or nominal target variables with only two values). Thus, some versions to work with a higher number of target values have been already proposed [1].

4.3.2 Extensions of Association Rule Mining Algorithms

The subgroup discovery problem has been also considered as an extension of association rule mining where the consequent of the rule is previously fixed to the target variable. In this sense, and thanks to the characteristics of the association rule mining algorithms that make possible to adapt them for the subgroup discovery task, many research studies were focused on versions of association rule mining approaches.

Apriori-SD [26] (see Algorithm 4.2) was proposed as an adaptation of the Apriori algorithm [3] for mining association rules. This adaptation to the subgroup discovery task includes a new post-processing mechanism, a new quality measure for the rules mined, and a probabilistic classification of the transactions. In order to make Apriori-SD appropriate for subgroup discovery, the Apriori algorithm is considered by taking in consideration only rules whose right-hand sides consist of the target variable value. In this algorithm some minimum threshold values

---

Algorithm 4.2 Pseudo-code of the Apriori-SD algorithm

Require: $\Omega, mSupport, mConf, k$ ▷ dataset Ω, minimum
      support and minimum confidence to accept a subgroup, and maximum number k of times that
      a record can be covered by the rule set
Ensure: S ▷ set S of the best solutions
 1: $RuleSet \leftarrow$ apply the Apriori algorithm by considering the thresholds $mSupport$ and $mConf$
 2: $RecordWeights \leftarrow$ initialize the weights associated to each record in Ω
 3: $MajorityClass \leftarrow$ take the target value that most frequently appear in Ω
 4: $S \leftarrow \emptyset$
 5: **repeat**
 6:       $BestRule \leftarrow$ take the rule from $RuleSet$ with the highest weighted relative accuracy value
 7:       $S = S \cup BestRule$
 8:       $RuleSet \leftarrow RuleSet \setminus BestRule$
 9:       Update the weights from $RecordWeights$ of those records covered by $BestRule$
10:       Remove from Ω those records covered more than k times
11: **until** $\Omega = \emptyset$ or $RuleSet = \emptyset$
12: $S \leftarrow S \cup MajorityClass$
13: **return** S

---

for the support (see Eq. (4.1)) and confidence (see Eq. (4.3)) quality measures are
considered so all the resulting rules should satisfy such thresholds (see line 1,
Algorithm 4.2). After that, the set of extracted rules are ordered according to the
weighted relative accuracy quality measure (see Eq. (4.8)) in a post-processing step,
incorporating a weighting scheme (each data record is weighted according to the
rules that it satisfies). Apriori-SD runs in an iterative fashion (see lines 5 to 11) so,
in each iteration, it selects the best rule according to the weighted relative accuracy
metric and weights associated to all the records covered by this rule are updated;
records are eliminated when they are covered more than k times (see line 10). Once
there is no more records in data or all the rules first obtained by Apriori [3] are used,
the algorithm returns the set S of rules (see line 13, Algorithm 4.2).

Mueller et al. [35] proposed the SD4TS (Subgroup Discovery for Test Selection) algorithm as an extension of Apriori-SD [26]. SD4TS (see Algorithm 4.3)
was proposed as a new approach to test selection based on the discovery of
subgroups of patients sharing the same optimal test, and present its application
to breast cancer diagnosis. The algorithm starts by producing a set of candidates
of size 1, that is, only one attribute-value is considered in each solution. Only
those solutions that satisfy a minimum support threshold value are considered as
$OptimizableCandidates$, and those k best solutions are saved in $TopCandidates$
(see line 2, Algorithm 4.3). The worst quality measure value from this set is also
taken as $minValue$. Then, from the $OptimizableCandidates$ set, those candidates
that do not satisfy the $minValue$ threshold are removed and an iterative process
is carried out from the resulting set (see lines 5 to 21, Algorithm 4.3). In this
iterative process, new candidates are generated from the $OptimizableCandidates$
set and saved as candidates or as best solutions depending on their quality measure
values. In situations were new solutions are saved in $TopCandidates$, then it
should be checked whether the maximum size of this set is achieved, removing

---

Algorithm 4.3 Pseudo-code of the SD4TS algorithm

Require: $\Omega, mSupport, k$ ▷ dataset Ω, minimum support and maximum number k of best subgroups to be produced

Ensure: $TopCandidates$ ▷ set of best solutions

1: $OptimizableCandidates \leftarrow$ any 1 attribute-value-pair that satisfy the $mSupport$ threshold
2: $TopCandidates \leftarrow$ best k subgroups (taking a quality measure) from $OptimizableCandidates$
3: $minValue \leftarrow$ worst quality measure value from the $TopCandidates$ set
4: Remove subgroups from $OptimizableCandidates$ that do not satisfy $minValue$
5: **while** $OptimizableCandidates \neq \emptyset$ **do**
6:    $c_1 \leftarrow$ take best candidate from $OptimizableCandidates$
7:    **for** $\forall c \in OptmizableCandidates$ **do**
8:       $c_{new} \leftarrow$ generate new candidates from c_1 and c
9:       **if** $support(c_{new}, \Omega) > mSupport$ and $qualityMeasure(c_{new}, \Omega) > minValue$ **then**
10:          $OptimizableCandidates \leftarrow c_{new}$
11:          **if** c_{new} is better than the worst from $TopCandidates$ **then**
12:             $TopCandidates \leftarrow c_{new}$
13:             **if** size of $TopCandidates$ is greater than k **then**
14:                Remove the worst solution from $TopCandidates$
15:             **end if**
16:             $minValue \leftarrow$ worst quality measure value from the $TopCandidates$ set
17:             Remove subgroups from $OptimizableCandidates$ that do not satisfy $minValue$
18:          **end if**
19:       **end if**
20:    **end for**
21: **end while**
22: **return** $TopCandidates$

---

the worst solution in such a situation (see lines 13 to 15, Algorithm 4.3). Finally, SD4TS returns the set of k best solutions found during its running, that is, the $TopCandidates$ set. It is important to remark that SD4TS was designed for a specific problem and, therefore, the quality measures were particularly designed for that problem.

SD-Map [5] is an exhaustive search method that depends on a minimum support threshold value to reduce the search space. In situations where the minimum support value is set to zero, then SD-Map performs an exhaustive search covering the whole unpruned search space. SD-Map uses the FP-Growth [24] algorithm, which differs from Apriori in its ability to reduce the database scans by considering a compressed representation of the database thanks to a data structure based on a traditional prefix-tree [42]. This structure represents each node by using an item and the frequency of the itemset denoted by the path from the root to that node. The resulting set includes only subgroups whose support is greater than the minimum predefined value. After that, a desired quality measure is computed for each of the solutions, and those subgroups that do not satisfy a minimum quality threshold are not considered.

In 2008, *H. Grosskreutz* [23], and similarly to SD-Map, proposed an algorithm for mining subgroups by considering a prefix-tree [42] to speed up the mining process. The main novelty of this proposal is the pruning of the search space by using optimistic estimates. Based on k subgroups already found for which it is known that all refinements s' of a subgroup s had a quality that is worse than that of all k subgroups found so far, then it is possible to prune that branch. The only requirement is an optimistic estimate for the refinements s' of s. In other words, given a subgroup s, an optimistic estimate is a function that provides a bound for the quality of every subgroup s' that is a refinement of s. One year later, the same authors proposed the Merge-SD algorithm [22], illustrating the importance of a good replacement of numerical attributes by nominal attributes, which can result in suboptimal results. Merge-SD was designed to prune a large part of the search space by exploiting constraints among the quality of subgroups ranging over overlapping intervals of the same numerical attribute.

SD-Map* [30] was proposed in 2016 as an algorithm that improves its predecessor SD-Map [5]. Whereas, SD-Map focuses exclusively on binary targets, SDMap* extends the employed numerical target concepts. In SD-Map*, a data structure based on a traditional prefix-tree [42] is considered, where the structure represents each node by using an item and two variables for the itemset denoted by the path from the root to that node. Such variables store the sum of the target values and the record count, enabling the computation of the mean target value of a set of records. Using these adaptations for numerical target variables, the case of a binary target (described for SD-MAP [5]) is considered as a special case if the value of the target is set to 1 for true target concepts, or set to 0 for false target concepts, respectively. Additionally, the statistics contained in the nodes of the resulting tree structures are used to compute not only the interestingness of subgroups, but also their optimistic estimates.

4.3.3 Evolutionary Algorithms

According to some authors [11], and due to the subgroup discovery task can be approached as a search and optimization problem, some authors have treated the problem from an evolutionary perspective. Additionally, evolutionary algorithms are very suitable to perform this task since they reflect well the interaction of variables in rule-learning processes and also provide a wide flexibility in the representation [43]. Analysing the specialized bibliography in this matter, almost all the existing evolutionary algorithms for mining subgroups are based on a "chromosome = subgroup" approach and the use of a single objective as fitness function to be optimized.

One of the first evolutionary algorithms in the subgroup discovery task was proposed in 2007 [17]. This algorithm, known as SDIGA (Subgroup Discovery Iterative Genetic Algorithm), is a genetic fuzzy system [45] that obtains fuzzy rules in disjunctive normal form. The fuzzy sets corresponding to the linguistic labels are defined by means of the corresponding membership functions, which can be specified either by the user or by means of a uniform partition with triangular membership functions (see Fig. 4.2). This type of rules allows to represent the knowledge in an explanatory and understandable.

Fig. 4.2 Example of fuzzy partition of a numerical variable considering five different linguistic labels

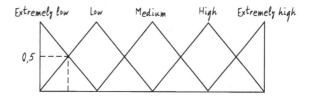

The SDIGA algorithm is run once for each different value in the target variable so, in the simplest problem (binary target variable), the algorithm runs two times. SDIGA represents the solutions in a fixed-length chromosome with a binary representation where a single bit is used for each feasible attribute-value pair. It should be highlighted that the target variable is not considered by the chromosome since all the individuals (subgroups or solutions to the problem) in the population are associated with the same target value. Analysing the chromosome, an active bit (a value of 1 for that bit) indicates that the corresponding attribute-value pair is used in the subgroup. In situations where an individual is represented with all the bits corresponding to a specific attribute with the value 1, or all of them contain the value 0, then the attribute has no relevance in the subgroup and such attribute is ignored. For a matter of clarification, let us consider a dataset comprising four attributes where the first attribute includes three different values, the second attribute comprises two different values, the third attribute two values, and the fourth attribute four values. Thus, the length of the chromosome is 11 and the sample solution (00100011000) represents the subgroup $Attribute_1 = value_3, Attribute_3 = value_2, Attribute_4 = value_1$. The second attribute is ignored since all their bits are 0.

The pseudo-code of the SDIGA algorithm is shown in Algorithm 4.4, which is an iterative procedure that is performed as many times as number target values exists in the problem. For each of the target values, SDIGA iterates (see lines 3 to 21, Algorithm 4.4) a number of times taking the best rule that is highly reliable (according a minimum confidence value) and covers any record of data that was not previously covered. SDIGA includes a genetic algorithm that runs a maximum number of generations (see lines 8 to 15, Algorithm 4.4) to obtain the best rules. This genetic algorithm uses a steady-state reproduction model in which the original population is only modified through the substitution of the worst individuals by individuals resulting from crossover and mutation. After that, a local search is carried out as a postprocessing phase (line 16, Algorithm 4.4) to increase the support of the subgroup throughout a hill-climbing process. Finally, and focusing on the evaluation process (lines 7 and 11, Algorithm 4.4), two different quality measures to be maximized are considered: support (general subgroups) and confidence (accurate subgroups). The function to evaluate each generated subgroup s in SDIGA is computed as $f(s) = (w_1 \times support(s) + w_2 \times confidence(s))/(w_1 + w_2)$, which is a weighted sum method that weights the two quality measures to be maximized. The determination of proper values for the weights, that is, w_1 and w_2, depends on the importance given by the users to each measure.

Algorithm 4.4 Pseudo-code of the SDIGA algorithm

Require: $\Omega, maxGenerations, mConfidence$ ▷ dataset, maximum number of generations to be considered and minimum confidence threshold

Ensure: *subgroups*

1: **for** $\forall t \in \Omega$ **do**   ▷ for each value in the target variable
2:      $subgroups \leftarrow \emptyset$
3:      **repeat**
4:           $population \leftarrow \emptyset$
5:           $number\_generations \leftarrow 0$
6:           Create a set of *subgroups* for the target t
7:           Evaluate each subgroup within *subgroups* by considering Ω
8:           **while** $number\_generations < maxGenerations$ **do**
9:                $parents \leftarrow$ select parents from *population*
10:               $offspring \leftarrow$ apply genetic operators to *parents*
11:               Evaluate each subgroup within *offspring* by considering Ω
12:               $bestSubgroup \leftarrow$ take the best subgroup from $\{population, offspring\}$
13:               $population \leftarrow$ take the best individuals from $\{population, offspring, bestSubgroup\}$
14:               $number\_generations + +$
15:           **end while**
16:           run a local search to improve *bestSubgroup*
17:           **if** $confidence(bestSubgroup, \Omega) \geq mConfidence$ and *bestSubgroup* satisfies uncovered records in Ω **then**
18:               $subgroups \leftarrow subgroups \cup bestSubgroup$
19:               mark in Ω the set of records covered by *bestSubgroup*
20:           **end if**
21:      **until** $confidence(bestSubgroup, \Omega) < mConfidence$ or *bestSubgroup* does not represent new records
22: **end for**
23: **return** *subgroups*

Another approach based on an iterative rule learning methodology is GAR-SD [37], which is able to operate with both continuous (using intervals) and discrete attributes. This algorithm represents the solutions (subgroups) as linear chromosomes that denote the values of the features that describe the subgroup for a specific target value. GAR-SD represents each individual as a linear chromosome where features defined in a continuous domains are represented by means of lower and upper interval values. Thus, the i-th feature defined in a continuous domain includes two values, that is, l_i and u_i. On the contrary, features defined in a discrete domain are represented by a single value within the chromosome. As a result, the length of a chromosome for a sample dataset comprising 4 numerical attributes and 6 discrete attributes is $4 \times 2 + 6 = 14$.

The GAR-SD algorithm (see Algorithm 4.5) is based on a traditional genetic algorithm that is run for each target value t in data. The algorithm works by taking the best subgroup that can be found by a genetic algorithm (see lines 8 to 15, Algorithm 4.5), considering a maximum threshold value α of records in Ω not covered yet by the resulting set of discovered subgroups. In the genetic algorithm, new subgroups are generated through the application of genetic operators. Solutions are evaluated (see line 7 and 11) according to a function that is a weighted

---

Algorithm 4.5 Pseudo-code of the GAR-SD algorithm

Require: $\Omega, maxGenerations, \alpha$ ▷ dataset, maximum number of generations to be considered
and maximum threshold value of non-covered records
Ensure: *subgroups*
 1: *subgroups* ← ∅
 2: **for** $\forall t \in \Omega$ **do** ▷ for each value in the target variable
 3:     **while** number of non covered records in Ω is lower than α **do**
 4:         S ← ∅
 5:         *number_generations* ← 0
 6:         Create a set S of subgroups for the target value t
 7:         Evaluate each subgroup within S by considering Ω
 8:         **while** *number_generations* < *maxGenerations* **do**
 9:             *parents* ←select parents from S
10:             *offspring* ←apply genetic operators to *parents*
11:             Evaluate each subgroup within *offspring* by considering Ω
12:             *bestSubgroup* ← take the best subgroup from {$S, offspring$}
13:             S ← take the best individuals from {$S, offspring, bestSubgroup$}
14:             *number_generations* + +
15:         **end while**
16:         Check the set of non covered records in Ω
17:         *subgroups* ← *subgroups* ∪ *bestSubgroup*
18:     **end while**
19: **end for**
20: **return** *subgroups*

---

aggregation function comprising different objectives. Some of these objectives are related to some quality measures for subgroups discovery: support, confidence and significance. On the contrary, other objectives are related to the variables that form the rule, e.g. the average amplitude of all the numeric intervals within the rule.

MESDIF (Multiobjective Evolutionary Subgroup DIscovery Fuzzy rules) [10] was proposed as one of the first genetic algorithms for mining subgroups that was based on a multi-objective methodology [14]. According to authors, the simultaneous optimization of different objectives at time is desirable in many situations, specially since many quality measures in subgroup discovery involve conflicting objectives (the improvement of one objective may lead to deterioration of another). Thus, there is not a single solution that optimizes all objectives simultaneously and the best trade-off solutions are required.

In MESDIF, different linguistic labels are assigned to continuous features and all the solutions are then represented as a fixed-length chromosome with a binary representation. An active bit (a value of 1 for that bit) indicates that the corresponding attribute-value (linguistic label) pair is used in the subgroup, whereas a 0 value indicates that the corresponding linguistic label or discrete value of the feature is not used in the subgroup. In situations where a specific attribute or feature has all its bits to 1, then it denotes that such attribute is not included in the subgroup description.

The methodology followed by MESDIF is inspired by SPEA2 [46] (see Algorithm 4.6), which is one of the most well-known multi-objective algorithms.

Algorithm 4.6 Pseudo-code of the MESDIF algorithm

Require: Ω, *population_size, elite_size* ▷ dataset and maximum sizes for population and elite
       population
Ensure: *subgroups*
 1: *subgroups* ← ∅
 2: **for** ∀t ∈ Ω **do**   ▷ for each value in the target variable
 3:     *population* ← ∅
 4:     *elite_population* ← ∅
 5:     *parents* ← ∅
 6:     *offspring* ← ∅
 7:     *population* ← generate(*population_size*)
 8:     Evaluate each subgroup within *population* by considering Ω
 9:     **while** termination condition is not reached **do**
10:         *elite_population* ← obtain the pareto front from {*population, elite_population*}
11:         **if** size(*elite_population*) > *elite_size* **then**
12:             *elite_population* ← reduce the set *elite_population* to the size of *elite_size*
13:         **else**
14:             **if** size(*elite_population*) < *elite_size* **then**
15:                 *elite_population* ← add new solutions to *elite_population* from
       population
16:             **end if**
17:         **end if**
18:         *parents* ←select parents from *population* by a binary tournament selector
19:         *offspring* ←apply genetic operators to *parents*
20:         Evaluate each subgroup within *population* by considering Ω
21:         *population* ← take the best individuals from {*population, offspring*}
22:     **end while**
23:     *subgroups* ← *subgroups* ∪ *elite_population*
24: **end for**
25: **return** *subgroups*

In order to quantify the quality of the solutions (subgroups), MESDIF considers
several quality measures (confidence, support, significance and unusualness) to
be optimized at time. A major feature of MESDIF is its ability to maintain the
best solutions found along the whole evolutionary process. To do so, it applies the
concept of elitism in the selection of the best subgroups by using a secondary or
elite population (lines 10 to 17, Algorithm 4.6). In order to preserve the diversity of
the set of solutions, a technique based on niches is considered, which analyses the
proximity in values of the objectives and an additional objective based on novelty
to promote subgroups which give information on uncovered data records.

Another multi-objective evolutionary algorithm based on fuzzy systems for
mining subgroups was proposed by *Carmona et al.* [9]. This algorithm, named
NMEEF-SD (Non-dominated Multi-objective Evolutionary algorithm for Extracting Fuzzy rules in Subgroup Discovery), is based on a multi-objective approach
to optimize a set of quality measures at once (confidence, support, sensitivity,
significance and unusualness). Unlike MESDIF, which was inspired by SPEA2 [46],
NMEEF-SD is based on the well-known NSGA-II [16] multi-objective algorithm.
A major feature of NMEEF-SD is its ability to represent solutions including

numerical features without the need for a previous discretization procedure. Here, numerical attributes are transformed into linguistic labels by means of a fuzzy system that contributes to the interpretability of the extracted subgroups. In order to represent solutions within the evolutionary process, NMEEF-SD uses a fixed-length chromosome representation of integer values without considering the target variable. Here, the i-th value within the chromosome denotes the i-th feature of the dataset, and the value represents the order of the discrete value (or linguistic label) except for zero values that denote unused attributes. As a matter of clarification, let us consider a dataset comprising four features. A sample solution represented as a linear chromosome of integers is (1013), illustrating a sample subgroup where the first attribute takes the first discrete value (or first linguistic label); the second attribute is not considered in the subgroup (it is a zero value); the third attribute takes the first discrete value (or first linguistic label) and, finally, the fourth attribute takes the third discrete value (or first linguistic label).

As described above, NMEEF-SD is based on the NSGA-II algorithm [16] and it evolves the set of individuals that are organized in fronts [14] based on the objectives to be optimized. No solution of a front can be said to be better than the others within the same front, but solutions from the first front are better than solutions from the second front, and so on. A major feature of NMEEF-SD is that it produces solutions by considering a biased initialization (see line 7, Algorithm 4.7), aiming at obtaining a set of initial solutions that represent a high number of records in data. Another interesting procedure proposed by NMEEF-SD is the re-initialization based on coverage (see line 23, Algorithm 4.7). Additionally, NMEEF-SD checks whether the pareto front is improving along the evolutionary process, analysing solutions within *pareto_front* and testing new covered records. The algorithm considers the pareto front is being improved if it covers at least one data record that was not covered by the previous *pareto_front*. In such situations where *pareto_front* does not evolve for more than 5% of the evolutionary process (quantified through the number of evaluations), then a re-initialization process is carried out by applying following steps: (1) non-repeated individuals of *pareto_front* are directly replicated in the new population; (2) the new population is completed with individuals generated through initialization based on coverage.

In 2012, *Rodriguez et al.* [41] proposed EDER-SD (Evolutionary Decision Rules for Subgroup Discovery). This evolutionary algorithm was designed to work in continuous domains (only numerical features are considered) to describe discrete target variables. EDER-SD represents each individual as a chromosome including two values (the lower and upper interval value) for each attribute, and an additional value (last position in the chromosome) to represent the target value. When the lower l_i and upper u_i bounds of the interval of the i-th attribute are equal to their respective maximum boundaries, the such attribute is denoted as irrelevant it will not appear in the resulting subgroup.

Similarly to previous works based on evolutionary algorithms for mining subgroups, EDER-SD was designed by means of an iterative rule learning methodology (see Algorithm 4.8), discovering the best subgroups in each iteration. In order to avoid subgroups that covers the same set of records in data, once a subgroup is

---

Algorithm 4.7 Pseudo-code of the NMEEF-SD algorithm

Require: Ω, $population\_size$, $elite\_size$ ▷ dataset and maximum sizes for population and elite
   population
Ensure: $subgroups$
 1: $subgroups \leftarrow \emptyset$
 2: **for** $\forall t \in \Omega$ **do**   ▷ for each value in the target variable
 3:     $population \leftarrow \emptyset$
 4:     $pareto\_front \leftarrow \emptyset$
 5:     $parents \leftarrow \emptyset$
 6:     $offspring \leftarrow \emptyset$
 7:     $population \leftarrow$ generate $population\_size$ with biased initialization
 8:     **while** termination condition is not reached **do**
 9:         $parents \leftarrow$ select parents from $population$ by a binary tournament selector
10:         $offspring \leftarrow$ apply genetic operators to $parents$
11:         Evaluate each subgroup within $population$ by considering Ω
12:         $population \leftarrow$ take the best individuals from $\{population, offspring\}$
13:         Generate all non-dominated fronts $F = (F_1, F_2, ...)$ from $population$
14:         Calculate the crowding-distance in any F_i
15:         $pareto\_front \leftarrow F_1$
16:         **if** $pareto\_front$ evolves **then**
17:             $i \leftarrow 2$
18:             **while** $|population| + |F_i| \leq population\_size$ **do**
19:                 $population \leftarrow population \cup F_i$
20:                 i++
21:             **end while**
22:         **else**
23:             Apply re-initialization based on coverage
24:         **end if**
25:     **end while**
26: **end for**
27: **return** $subgroups$

---

obtained, those records that were already covered by other subgroups are penalised
(lines 16 to 20, Algorithm 4.8). Hence, when analysing the records in the first
iteration of the evolutionary procedure, all data records are equally considered,
whereas in posteriori iterations, previously covered transactions are penalised and
they are considered as less important for the calculation of the importance of the
subgroup it is. In this evaluation process, a wide number of the metrics already
used in subgroup discovery are considered (unusualness, confidence, sensitivity and
significance).

One of the last evolutionary proposals proposed so far in the subgroup discovery
task dates from 2014. This algorithm, known as CGBA-SD (Comprehensible
Grammar-Based Algorithm for Subgroup Discovery) [32], was the first subgroup
discovery algorithm based on grammar guided genetic programming [33]. Similarly
to any of the existing evolutionary algorithms in the subgroup discovery field,
CGBA-SD follows an iterative methodology where subgroups of a specific target
feature are mined in each running. In CGBA-SD, unlike previously described
algorithms that represented the solutions by a linear chromosome, solutions are

---

Algorithm 4.8 Pseudo-code of the EDER-SD algorithm

Require: $\Omega, maxGenerations$ ▷ dataset and number of generations to be considered
Ensure: *subgroups*
1: *subgroups* ← ∅
2: **for** $\forall t \in \Omega$ **do**   ▷ for each value in the target variable
3:      **while** ∃ a transaction for t not covered yet **do**
4:         S ← ∅
5:         *number_generations* ← 0
6:         Create a set S of subgroups for the target value t
7:         Evaluate each subgroup within S by considering Ω
8:         **while** *number_generations* < *maxGenerations* **do**
9:            *parents* ←select parents from S
10:        *offspring* ←apply genetic operators to *parents*
11:        Evaluate each subgroup within *offspring* by considering Ω
12:        *bestSubgroup* ← take the best subgroup from $\{S, offspring\}$
13:        S ← take the best individuals from $\{S, offspring, bestSubgroup\}$
14:        *number_generations* + +
15:      **end while**
16:      **for** $\forall records \in \Omega$ **do**
17:        **if** *record* is covered by *bestSubgroup* **then**
18:           Penalize *record* using a threshold
19:        **end if**
20:      **end for**
21:      *subgroups* ← *subgroups* ∪ *bestSubgroup*
22:     **end while**
23: **end for**
24: **return** *subgroups*

---

represented as a syntax-tree encoded through a context-free grammar (see Fig. 4.3). For a better understanding of the concept of context-free grammar, it is defined as a four-tuple $(\Sigma_N, \Sigma_T, P, S)$ where Σ_T represents the alphabet of terminal symbols and Σ_N the alphabet of non-terminal symbols. Noted that they have no common elements, i.e. $\Sigma_N \cap \Sigma_T = \emptyset$. In order to encode a solution using a G3P approach, a number of production rules from the set P are applied beginning from the start symbol S. A production rule is defined as $\alpha \rightarrow \beta$ where $\alpha \in \Sigma_N$, and $\beta \in \{\Sigma_T \cup \Sigma_N\}^*$. To obtain solutions, a number of production rules is applied from the set P, obtaining a derivation syntax tree for each solution, where internal nodes contain only non-terminal symbols, and leaves contain only terminal symbols. Considering this grammar G (see Fig. 4.3), the following language is obtained $L(G) = \{(AND \ Condition)^n \ Condition \rightarrow \ Target : n \geq 0\}$. Thanks to this grammar, it is possible to mine any subgroup containing either numerical or discrete features so it is runnable on each specific application domain or problem.

CGBA-SD is an iterative process (see Algorithm 4.9) that is repeated as many times as number of distinct values included in the target feature. In each generation of CGBA-SD (see lines 7 to 14, Algorithm 4.9), those solutions (subgroups) exceeding a previously specified minimum confidence value are selected to be included in an external or elite population (see line 12, Algorithm 4.9). This

$G = (\Sigma_N, \Sigma_T, P, S)$ with:

S = Subgroup
Σ_N = {Subgroup, Antecedent, Target, Nominal_Condition, Numerical_Condition }
Σ_T = {'AND', 'Attribute', 'Target_value', '=', 'IN','Min_value', 'Max_value' }
P = {Subgroup ::= Antecedent, Class ;
         Antecedent ::= Nominal_Condition | Numerical_Condition |
                        'AND', Nominal_Condition, Antecedent |
                        'AND', Numerical_Condition, Antecedent ;
         Nominal_Condition ::= 'Attribute', '=', 'value' ;
         Numerical_Condition ::= 'Attribute', 'IN', 'Min_value', 'Max_value' ;
         Target ::= 'target', '=', 'Target_value'; }

Fig. 4.3 Context-free grammar used to encode individuals in the CGBA-SD algorithm

---

Algorithm 4.9 Pseudo-code of the CGBA-SD algorithm

Require: $\Omega, maxGenerations, mConfidence$ ▷ dataset, number of generations to be
     considered and minimum confidence threshold
Ensure: *subgroups*
1: **for** $\forall t \in \Omega$ **do**   ▷ for each value in the target variable
2:     *subgroups* $\leftarrow \emptyset$
3:     *population* $\leftarrow \emptyset$
4:     *number_generations* $\leftarrow 0$
5:     Create a set *population* of subgroups for the target value t
6:     Evaluate each subgroup within *population* by considering Ω
7:     **while** *number_generations* $< maxGenerations$ **do**
8:        *parents* $\leftarrow$ select parents from *population*
9:        *offspring* $\leftarrow$ apply genetic operators to *parents*
10:       Evaluate each subgroup within *offspring* by considering Ω
11:       *population* $\leftarrow$ take the best subgroups from {*population, offspring,subgroups*}
12:       *subgroups* $\leftarrow$ best individuals from *population* that overcome $mConfidence$
13:       *number_generations* $++$
14:     **end while**
15:     optimize bounds of the numerical intervals from *subgroups* **return** *subgroups*
16: **end for**

---

selection procedure works as an elitist selection, allowing the best found solutions to be considered in the next generation. In each iteration, CGBA-SD generates new individuals (subgroups) through genetic operators (see line 9, Algorithm 4.9) and, taking the dataset Ω, each new individual s is evaluated according to a function that is defined as $f(s) = support(x, \Omega) \times confidence(s, \Omega)$. Thus, the goal of CGBA-SD is to mine reliable subgroups that cover a high percentage of data records for each target value. Finally, a major feature of CGBA-SD is its procedure for optimizing intervals in numerical attributes. This procedure, which is based on the well-known hill climbing optimization method, increases or decreases the width of the intervals in order to discover high quality intervals that satisfy more data records or which prompt more reliable subgroups. Thus, once the algorithm has discovered subgroups for each of the values of the target feature, CGBA-SD runs the

---

Algorithm 4.10 Pseudo-code of the Map procedure for the MEFASD-BD algorithm

---

Require: Ω, $population\_size$, $mConfidence$ ▷ dataset, maximum population size and
     minimum confidence threshold
Ensure: $subgroups$
 1: $subgroups \leftarrow \emptyset$
 2: **for** $\forall t \in \Omega$ **do**   ▷ for each value in the target variable
 3:      $population \leftarrow \emptyset$
 4:      $parents \leftarrow \emptyset$
 5:      $offspring \leftarrow \emptyset$
 6:      $number\_generations \leftarrow 0$
 7:      $population \leftarrow$ generate $population\_size$ individuals for the target t
 8:      **while** $number\_generations < maxGenerations$ **do**
 9:          $parents \leftarrow$ select parents from $population$
10:          $offspring \leftarrow$ apply genetic operators to $parents$
11:          Evaluate each subgroup within $population$ by considering Ω
12:          $population \leftarrow$ take the best individuals from $\{population, offspring\}$
13:          Generate all non-dominated fronts $F = (F_1, F_2, ...)$ from $population$
14:          $pareto\_front \leftarrow F_1$
15:          $population \leftarrow$ take solutions from the fronts in order till $population$ is completed
16:      **end while**
17:      $subgroups \leftarrow$ individuals from pareto front with confidence value above $mConfidence$
18: **end for**
19: **return** $subgroups$

---

aforementioned final procedure to optimize the intervals of the numerical conditions
(see line 15, Algorithm 4.9).

4.3.4   Big Data Approaches

Data science appears with the objective of extracting knowledge from large datasets,
but new technologies are required when extremely large datasets are analysed.
One of the most important and well-known emerging technologies for parallel
processing and intensive computing is MapReduce [15]. In recent years, there
is a growing interest in the development of new algorithms using MapReduce
for big data problems and one of the promising areas is subgroup discovery.
In this regard, MEFASD-BD (Multi-objective Evolutionary Fuzzy Algorithm for
Subgroup Discovery in Big Data environments) [40] is a recent algorithm for mining
subgroups in big data. MEFASD-BD takes the basic ideas of the NMEEF-SD
algorithm [9], which was described in the previous section (see Algorithm 4.7).

MEFASD-BD works by first dividing the dataset into a number of partitions and
each of these partition is loaded in a map procedure (see Algorithm 4.10). In each
of these map procedures, an evolutionary process is carried out and it is run as many
times as number of distinct values of the target variable. In this initial procedure,
all the obtained subgroups are non-dominated with respect to unusualness (see
Eq. (4.8)) and sensitivity (see Eq. (4.11)) quality measures. Since a map procedure

---

Algorithm 4.11 Pseudo-code of the Reduce procedure for the MEFASD-BD algorithm

---

Require: *subgroups*   ▷ set of subgroups obtained in the map procedure
Ensure: *subgroups*
 1: Order *subgroups* by global unusualness
 2: Apply token competition operator
 3: Delete solutions without tokens from *subgroups*
 4: **return** *subgroups*

---

is carried out for each partition of the original dataset, independent subgroups are obtained for each partition. The map procedure is carried for each value t of the target variable (see lines 2 to 18, Algorithm 4.10), where set of solutions is first is generated and an evolutionary process controlled by a number of evaluations is then run.

Once the map procedure is done, MEFASD-BD has obtained a set of rules for each map considering only its own subset of data instead of the whole dataset. Thus, it is required to compute the quality of each subgroup with regard to the whole dataset. At this point, and once the obtained subgroups are analysing through the whole dataset, the confidence values for each subgroup may vary so they are analysed and those subgroups whose global confidence is under a predefined threshold are discarded. After that, a final stage is performed (see Algorithm 4.11), which is relative to reduce phase that first combines all the sets of subgroups obtained by the map procedure. Next, a sorting based on the global unusualness computed in the previous stage is performed by applying token competition where each data record is a token and all the subgroups must compete for this token. When an individual (subgroup) covers the provided record, then this record will not be satisfied by other individuals. This process is applied for each individual in such a way that individual with the highest global unusualness take as many tokens as possible.

The search space in the subgroup discovery task has been reduced by means of heuristic algorithms [43]. Nevertheless, although these techniques have achieved to reduce the search space, many interesting solutions could be neglected. In this regard, exhaustive search proposals are required even when they are computationally expensive for large datasets. In a recent research study, *Padillo et al.* [39] proposed two MapReduce solutions, AprioriKSD-OE (AprioriK-Subgroup Discovery Optimistic Estimate) and PFP-SD-OE (Parallel FP-Growth Subgroup Discovery Optimistic Estimate), to work on big data that are based on optimistic estimates to prune the search space achieving a better runtime without losing any accuracy in the results.

AprioriKSD-OE [39] is an exhaustive search algorithm based on the well-known Apriori [3] algorithm. A major characteristic of this algorithm is it obtains subgroups on any kind of target value, changing the quality metric depending on the target type (binary, discrete or numerical). Since this algorithm is based on MapReduce, it uses three different processes: driver, mapper and reducer. In the first step, the driver reads the database from disk, split the database into subsets, and calculates the maximum number of attributes in each subgroup (maximum number of iterations that will be performed). In the second step, for each iteration l, a MapReduce phase is needed (see Algorithms 4.12 and 4.13). In this regard, a different mapper is run for each subset of data and the results are collected in a reducer phase. Thus, this second step is split in two different subprocesses, that is, the Map and Reduce processes. In the Map phase (see Algorithm 4.12), each of generated mappers are responsible for mining the complete set of subgroups of size l (depending on the iteration carried out) for its sub-database. Here, a set of $\langle k, v \rangle$ pairs are produced, where k denotes the set of attributes and v is the value for such set. The v value depends on the type of target so, for example, when working with numerical target variables the v value will be comprised of the average value for the target, the number of records satisfied by k and the average value for the whole dataset. In the Reduce phase (see Algorithm 4.13), the overall count for each subgroup is performed returns to the driver only those subgroups with a quality greater than a threshold. Finally, a third step is carried out after the MapReduce phase. Here, the driver collects the j best subgroups of size l, which are joined with the best discovered subgroups up to now. The best optimistic estimate of the subgroups of size l is compared with the worst quality of the subgroups of whatever size. In case that it is lower, then the process could be stopped since any refinement could not be better than the previously obtained. In another case, l is incremented and it returns to the second step. This iterative process is carried out till l reaches the maximum value.

PFP-SD-OE [38] is a MapReduce solution for mining subgroups that is based on a parallel version of FPGrowth [24]. The dataset is read in a distributed way enabling a faster computing and creating a set of independent FP-Trees [24] to represent the current dataset. The number of FP-Trees is determined by the number of different single values for each attribute. Then, the FP-Trees are mined in a parallel way and each computer node analyse a single tree. Moreover, optimistic estimates are considered to reduce the search process on different types of targets (binary, discrete and numerical). PFP-SD-OE includes two different MapReduce processes that are performed in a number of steps. In the first step, the database is read from disk and it is split into subsets. In the second step, a parallel counting is carried out (see Algorithm 4.14) where the frequency for each value of each attribute is calculated to produce a final list called F-list (see Algorithm 4.15). This

---

Algorithm 4.12 Pseudo-code of the Map procedure for the AprioriKSD-OE algorithm

---

Require: $l, record$ ▷ length l of the subgroups to be obtained and data record
 1: $subgroups \leftarrow$ create all the subgroups of size l from $record$
 2: **for** $\forall subgroup \in subgroups$ **do**   ▷ for each of the produced subgroups
 3:    $info \leftarrow$ Generate information to be assigned depending on the type of target
 4:    Emit the pair $\langle subgroup, info \rangle$ ▷ $\langle k, v \rangle$ pairs
 5: **end for**

Algorithm 4.13 Pseudo-code of the Reduce procedure for the AprioriKSD-OE algorithm

---

Require: $subgroup, threshold\_quality$ ⊳ subgroup to be computed and minimum threshold value
1: $globalInfo \leftarrow 0$
2: **for** $\forall info \in subgroup$ **do**   ⊳ analyse all the v values (info) obtained in the Map phase
3:     $globalInfo \leftarrow globalInfo + info$
4:     **if** $globalInfo \geq threshold\_quality$ **then**
5:         Emit the pair $\langle subgroup, globalInfo \rangle$ ⊳ $\langle k, v \rangle$ pairs
6:     **end if**
7: **end for**

Algorithm 4.14 Pseudo-code of the Map procedure for the parallel counting of the PFP-SD-OE algorithm

---

Require: $record$
1: **for** $\forall value \in record$ **do**   ⊳ for each value in the record
2:     Emit the pair $\langle value, 1 \rangle$ ⊳ $\langle k, v \rangle$ pairs
3: **end for**

Algorithm 4.15 Pseudo-code of the Reduce procedure for the parallel counting of the PFP-SD-OE algorithm

---

Require: k ⊳ key value from each $\langle k, v \rangle$ pair obtained in the Map phase
1: $globalFrequency \leftarrow 0$
2: **for** $\forall v_i \in v$ **do**
3:     $globalFrequency \leftarrow globalFrequency + v_i$
4: **end for**
5: $F - List.append(k, globalFrequency)$

---

step requires one read of the dataset, and hence, a whole MapReduce process is required. After this MapReduce process is performed, the resulting F-list is sorted by its frequency values and, then, a series of FP-Trees are created by following the procedure described in for a parallel FP-Growth [31].

References

1. T. Abudawood, P. Flach, Evaluation measures for multi-class subgroup discovery, in *Machine Learning and Knowledge Discovery in Databases*, ed. by W. Buntine, M. Grobelnik, D. Mladenić, J. Shawe-Taylor. Lecture Notes in Computer Science, vol. 5781 (Springer, Berlin, 2009), pp. 35–50
2. C.C. Aggarwal, J. Han, *Frequent Pattern Mining* (Springer International Publishing, Cham, 2014)
3. R. Agrawal, T. Imielinski, A.N. Swami, Mining association rules between sets of items in large databases, in *Proceedings of the 1993 ACM SIGMOD International Conference on Management of Data (SIGMOD Conference '93)*, Washington, DC, pp. 207–216 (1993)
4. M. Atzmueller, Subgroup discovery - advanced review. WIREs Data Min. Knowl. Discovery **5**, 35–49 (2015)

5. M. Atzmueller, F. Puppe, SD-Map – a fast algorithm for exhaustive subgroup discovery, in *Proceedings of the 10th European Symposium on Principles of Data Mining and Knowledge Discovery (PKDD '06)*, Berlin, pp. 6–17 (2006)

6. M. Atzmuller, F. Puppe, H.P. Buscher, Towards knowledge-intensive subgroup discovery, in *Proceedings of the Lernen-Wissensentdeckung-Adaptivitat-Fachgruppe Maschinelles Lernen (LWA-04)*, Berlin, pp. 111–117, October 2004

7. S.D. Bay, M.J. Pazzani, Detecting group differences: mining contrast sets. Data Min. Knowl. Disc. **5**(3), 213–246 (2001)

8. O. Bousquet, U. Luxburg, G. Ratsch, *Advanced Lectures On Machine Learning* (Springer, Berlin, 2004)

9. C.J. Carmona, P. González, M.J. del Jesus, F. Herrera, NMEEF-SD: non-dominated multiobjective evolutionary algorithm for extracting fuzzy rules in subgroup discovery. IEEE Trans. Fuzzy Syst. **18**(5), 958–970 (2010)

10. C.J. Carmona, P. González, M.J. del Jesus, M. Navío-Acosta, L. Jiménez-Trevino, Evolutionary fuzzy rule extraction for subgroup discovery in a psychiatric emergency department. Soft Comput. **15**(12), 2435–2448 (2011)

11. C.J. Carmona, P. González, M.J. del Jesus, F. Herrera, Overview on evolutionary subgroup discovery: analysis of the suitability and potential of the search performed by evolutionary algorithms. Wiley Interdiscip. Rev. Data Min. Knowl. Disc. **4**(2), 87–103 (2014)

12. C.J. Carmona, M.J. del Jesus, F. Herrera, A unifying analysis for the supervised descriptive rule discovery via the weighted relative accuracy. Knowl. Based Syst. **139**, 89–100 (2018)

13. P. Clark, T. Niblett, The CN2 induction algorithm. Mach. Learn. **3**(4), 261–283 (1989)

14. C.A. Coello, G.B. Lamont, D.A. Van Veldhuizen, *Evolutionary Algorithms for Solving Multi-Objective Problems (Genetic and Evolutionary Computation)* (Springer, New York, 2006)

15. J. Dean, S. Ghemawat, MapReduce: simplified data processing on large clusters. Commun. ACM **51**(1), 107–113 (2008)

16. K. Deb, A. Pratap, S. Agrawal, T. Meyarivan, A fast elitist multi-objective genetic algorithm: NSGA-II. IEEE Trans. Evol. Comput. **6**, 182–197 (2000)

17. M.J. del Jesus, P. Gonzalez, F. Herrera, M. Mesonero, Evolutionary fuzzy rule induction process for subgroup discovery: a case study in marketing. IEEE Trans. Fuzzy Syst. **15**(4), 578–592 (2007)

18. G. Dong, J. Bailey (eds.), *Contrast Data Mining: Concepts, Algorithms, and Applications* (CRC Press, Boca Raton, 2013)

19. W. Duivesteijn, A.J. Knobbe, Exploiting false discoveries - statistical validation of patterns and quality measures in subgroup discovery, in *Proceedings of the 11th IEEE International Conference on Data Mining (ICDM 2011)*, Vancouver, BC, pp. 151–160, December 2011

20. D. Gamberger, N. Lavrac, Expert-guided subgroup discovery: methodology and application. J. Artif. Intell. Res. **17**(1), 501–527 (2002)

21. A.M. García-Vico, C.J. Carmona, D. Martín, M. García-Borroto, M.J. del Jesus, An overview of emerging pattern mining in supervised descriptive rule discovery: taxonomy, empirical study, trends and prospects. Wiley Interdiscip. Rev. Data Min. Knowl. Disc. **8**(1) (2018)

22. H. Grosskreutz, S. Rüping, On subgroup discovery in numerical domains. Data Min. Knowl. Disc. **19**(2), 210–226 (2009)

23. H. Grosskreutz, S. Rüping, S. Wrobel, Tight optimistic estimates for fast subgroup discovery, in *Proceedings of the European Conference on Machine Learning and Principles and Practice of Knowledge Discovery in Databases (ECML/PKDD 08)*, Antwerp, pp. 440–456, September 2008

24. J. Han, J. Pei, Y. Yin, R. Mao, Mining frequent patterns without candidate generation: a frequent-pattern tree approach. Data Min. Knowl. Disc. **8**, 53–87 (2004)

25. F. Herrera, C.J. Carmona, P. González, M.J. del Jesus, An overview on subgroup discovery: foundations and applications. Knowl. Inf. Syst. **29**(3), 495–525 (2011)

26. B. Kavšek, N. Lavrač, APRIORI-SD: adapting association rule learning to subgroup discovery. Appl. Artif. Intell. **20**(7), 543–583 (2006)

27. W. Kloesgen, M. May, Census data mining - an application, in *Proceedings of the 6th European Conference on Principles and Practice of Knowledge Discovery in Databases (PKDD 2002)*, Helsinki (Springer, London, 2002), pp. 733–739
28. W. Klösgen, Explora: a multipattern and multistrategy discovery assistant, in *Advances in Knowledge Discovery and Data Mining*, ed. by U.M. Fayyad, G. Piatetsky-Shapiro, P. Smyth, R. Uthurusamy (American Association for Artificial Intelligence, Menlo Park, 1996), pp. 249–271
29. N. Lavrač, B. Kavšek, P. Flach, L. Todorovski, Subgroup discovery with CN2-SD. J Mach Learn Res **5**, 153–188 (2004)
30. F. Lemmerich, M. Atzmueller, F. Puppe, Fast exhaustive subgroup discovery with numerical target concepts. Data Min. Knowl. Disc. **30**(3), 711–762 (2016)
31. H. Li, Y. Wang, D. Zhang, M. Zhang, E.Y. Chang, PFP: parallel FP-growth for query recommendation, in *Proceedings of the 2008 ACM Conference on Recommender Systems*, Lausanne, October 2008 (ACM, New York, 2008), pp. 107–114
32. J.M. Luna, J.R. Romero, C. Romero, S. Ventura, On the use of genetic programming for mining comprehensible rules in subgroup discovery. IEEE Trans. Cybern. **44**(12), 2329–2341 (2014)
33. R. McKay, N. Hoai, P. Whigham, Y. Shan, M. O'Neill, Grammar-based Genetic Programming: a survey. Genet. Program. Evolvable Mach. **11**, 365–396 (2010)
34. T.M. Mitchell, *Machine Learning*. McGraw Hill Series in Computer Science (McGraw-Hill, Maidenhead, 1997)
35. M. Mueller, R. Rosales, H. Steck, S. Krishnan, B. Rao, S. Kramer, Subgroup discovery for test selection: a novel approach and its application to breast cancer diagnosis, in *Advances in Intelligent Data Analysis VIII*, ed. by N. Adams, C. Robardet, A. Siebes, J.F. Boulicaut. Lecture Notes in Computer Science, vol. 5772 (Springer, Berlin, 2009), pp. 119–130
36. P.K. Novak, N. Lavrač, G.I. Webb, Supervised descriptive rule discovery: a unifying survey of contrast set, emerging pattern and subgroup mining. J. Mach. Learn. Res. **10**, 377–403 (2009)
37. V. Pachón, J. Mata, J.L. Domínguez, M.J. Maña, A multi-objective evolutionary approach for subgroup discovery, in *Proceedings of the 5th International Conference on Hybrid Artificial Intelligence Systems (HAIS 2010)*, San Sebastian (Springer, Berlin, 2010), pp. 271–278
38. F. Padillo, J.M. Luna, S. Ventura, Subgroup discovery on big data: exhaustive methodologies using map-reduce, in *Proceedings of the 2016 IEEE Trustcom/BigDataSE/ISPA*, Tianjin (IEEE, Piscataway, 2016), pp. 1684–1691
39. F. Padillo, J.M. Luna, S. Ventura, Exhaustive search algorithms to mine subgroups on big data using apache spark. Prog. Artif. Intell. **6**(2), 145–158 (2017)
40. F. Pulgar-Rubio, A.J. Rivera-Rivas, M.D. Pérez-Godoy, P. González, C.J. Carmona, M.J. del Jesus, MEFASD-BD: multi-objective evolutionary fuzzy algorithm for subgroup discovery in big data environments - a mapreduce solution. Knowl. Based Syst. **117**, 70–78 (2017)
41. D. Rodriguez, R. Ruiz, J.C. Riquelme, J.S. Aguilar-Ruiz. Searching for rules to detect defective modules: a subgroup discovery approach. Inf. Sci. **191**, 14–30 (2012)
42. P.N. Tan, M. Steinbach, V. Kumar, *Introduction to Data Mining* (Addison Wesley, Boston, 2005)
43. S. Ventura, J.M. Luna, *Pattern Mining with Evolutionary Algorithms* (Springer International Publishing, Cham, 2016)
44. S. Wrobel, An algorithm for multi-relational discovery of subgroups, in *Proceedings of the 1st European Symposium on Principles of Data Mining and Knowledge Discovery (PKDD '97)*, London (Springer, Berlin, 1997), pp. 78–87
45. L.A. Zadeh, The concept of a linguistic variable and its application to approximate reasoning I,II,III. Inf. Sci. **8–9**, 199–249, 301–357, 43–80 (1975)
46. E. Zitzler, M. Laumanns, L. Thiele, SPEA2: improving the strength pareto evolutionary algorithm for multiobjective optimization, in *Proceedings of the 2001 conference on Evolutionary Methods for Design, Optimisation and Control with Application to Industrial Problems (EUROGEN 2001)*, Athens, pp. 95–100 (2001)

Chapter 5
Class Association Rules

Abstract Association rule mining is the most well-known task to perform data descriptions. This task aims at extracting useful an unexpected co-occurrences among items in data. Even when this task was generally defined for mining any type of association, it is sometimes desiderable the extraction of co-occurrences between a set of items and a specific target variable of interest. Thus, the general task is transformed into a more specific one, class association rule mining. This chapter introduces the class association rule mining problem and describes the main differences with regard to association rule mining. Then, the most widely used metrics in this field are analysed and a detailed description is also performed to denote the importance of using these associations. Finally, the most important approaches in this field are analysed, which are categorized into two different aims: predictive and descriptive.

5.1 Introduction

Data analysis is concerned with the development of methods and techniques for making sense of data [12]. Here, the key element is the pattern, representing any type of homogeneity in data and serving as a good data descriptor [1]. Sometimes, the insights extracted by a single pattern might not be enough and a more descriptive analysis is required. In this regard, *Agrawal et al.* [2] proposed the concept of association rules as a way of describing co-occurrences among sets of items within a pattern of potential interest.

Let P be a pattern defined as a subset of items $I = \{i_1, i_2, ..., i_n\}$ in a dataset Ω, i.e. $\{P = \{i_j, ..., i_k\} \subseteq I : 1 \leq j, k \leq n\}$. Let us also consider X and Y subsets of the pattern P, i.e. $X \subset P \subseteq I$ and $Y = P \setminus X$; or also $Y \subset P \subseteq I$ and $X = P \setminus Y$. An association rule [45] is formally defined as an implication of the form $X \rightarrow Y$ where X denotes the antecedent of the rule and Y the consequent. Here, it is satisfied that both X and Y have no common item, i.e. $X \cap Y = \emptyset$. The meaning of an association rule is that if the antecedent X is satisfied, then it is highly probable that the consequent Y is also satisfied. The mining of association rules was primarily focused on the market basket analysis [7], finding associations of items that are

© Springer Nature Switzerland AG 2018
S. Ventura, J. M. Luna, *Supervised Descriptive Pattern Mining*,
https://doi.org/10.1007/978-3-319-98140-6_5

generally bought together. Nowadays, however, the growing interest in descriptive analytics has given rise to the application of such associations to diverse fields. As an example, it has been used to identify complications of cerebral infarction in patients with atrial fibrillation [15], identifying a series of principal factors that include the age (older than 63 years old) and the hypertension in medical history. Another peculiar example came from NASA's satellites where, after analysing tons characteristics of the solar storms on the Sun and geomagnetic events around the Earth,[1] it was discovered that the strongest geomagnetic effects were obtained at approximately 2–3 h after the solar storm.

Association rule mining is a really challenging task since its process is computationally expensive and requires a large amount of memory [29]. Considering a dataset containing k items, i.e. $I = \{i_1, i_2, ..., i_k\}$, $\binom{k}{2}$ itemsets of length 2 can be produced and each of these itemsets produce $2^2 - 2 = 2$ different association rules. In the same dataset, there are $\binom{k}{3}$ itemsets of length 3, producing $2^3 - 2 = 6$ association rules. Thus, in general, we can determines that for a dataset comprising k singletons, a total of $3^k - 2^{k+1} + 1$ different association rules might be generated. As previously stated, the mining of association rules was first proposed as the extraction of implications of the form $X \rightarrow Y$ where X and Y may include any number of items with the only restriction that $X \cap Y = \emptyset$. Nowadays, however, there is a rising interest in using association rules for classification tasks [22]. In this sense, the goal is to mine association rules only including a specific attribute in the consequent, i.e. $X \rightarrow y$, X being a subset of items in data $X \subseteq I$ and y defined as the target variable or class attribute. These association rules are known as class association rules (CARs) and, among their various advantages for descriptive aims, e.g. they are easily understandable by the expert, they are defined in a search space that is much smaller than the one of general association rules. Given a dataset containing k items and a target variable comprising t different values, the total number of CARs that can be mined is $(2^k - 1) \times t$.

Since CARs were first introduced in 1998 by *Liu et al.* [22], many different algorithms have been proposed by different researchers. Such algorithms are mainly divided into two different tasks, that is, descriptive and predictive. Regarding descriptive tasks, most existing algorithms are modified versions of those for mining general association rules. Those algorithms based on two steps (one for mining frequent patterns and another for extracting association rules from the previous mined patterns) work in a similar fashion when CARs are desired. The frequent pattern mining process is carried out on the set of items by always including the target variable. Then, the extraction of CARs is easily performed since the consequent is always the target variable. On the contrary, those algorithms based on a single step (mostly approaches based on evolutionary computation) just need to calculate the frequencies of the patterns based on the premise of the target variable.

---

[1]K. Borne. *Association Rule Mining - Not Your Typical Data Science Algorithm.* April, 2014 - available at https://mapr.com/blog/association-rule-mining-not-your-typical-data-science-algorithm/.

As for predictive tasks, all the existing algorithms are based on two different steps, one for mining interesting CARs, and a second step for constructing an accurate classifier based on the previously mined rules.

Finally, it is important to remark the importance of CARs in numerous problems such as traffic load prediction [49], analysis of students' performance on learning management systems [28], and gene expression classification [17], among others.

5.2 Task Definition

The principal element in association rule mining is the pattern P, which is defined as a subset of the whole set of items $I = \{i_1, i_2, ..., i_l\}$ in a dataset Ω, i.e. $P = \{i_j, ..., i_k\} \subseteq I, 1 \leq j, k \leq l$. Taking the pattern P as the principal component, an association rule is then formally defined as an implication of the type $X \rightarrow Y$ that is formed from P in such a way that the antecedent X is defined as $X \subset P$, whereas the consequent Y is denoted as $Y = P \setminus X$. Here, it is satisfied that both X and Y have no common item, i.e. $X \cap Y = \emptyset$. The meaning of an association rule is that if the antecedent X is satisfied, then it is highly probable that the consequent Y is also satisfied.

The resulting set of association rules that is produced when applying any association rule mining algorithm can be decomposed into a more specific set including rules with a specific attribute in the consequent. These rules, known as class association rules (CARs) are formally defined as implications of the form $X \rightarrow y$, X being a subset of items in data $X \subseteq I$ and y defined as the target variable or class attribute. This subset of association rules may include a total of $(2^k - 1) \times t$ solutions for a dataset containing k items and a target variable comprising t different values. Thus, the resulting set of solutions is much smaller than the one of general association rules, which comprises up to $3^k - 2^{k+1} + 1$ solutions for a dataset comprising k items. Since its definition [22], CARs have gained an increasing attention due to their understandability and many research studies have considered these rules not only for descriptive purposes [28] but also to form accurate classifiers [22].

Finally, focusing on the quality measures defined to quantify the importance of each discovered association rule, a series of metrics [45] have been proposed in literature. Since CARs are specific cases of general association rules, all the existing metrics can be used for both types of rules. In general, any quality measure is based on the analysis of the statistical properties of data [5, 41] and they consider the number of data records satisfied by the rule (denoted as n_{xy}), the antecedent (n_x), and the consequent (n_y). Table 5.1(a) shows the relationships between these three values (n_{xy}, n_x and n_y) for a dataset Ω comprising $|\Omega|$ records. All these values (n_{xy}, n_x and n_y) can also be represented as relative frequencies (see Table 5.1(b)), by considering the number n of records in data, in per unit basis: $p_{xy} = n_{xy}/n$; $p_{x\overline{y}} = n_{x\overline{y}}/n$; $p_{\overline{x}y} = n_{\overline{x}y}/n$; $p_{\overline{xy}} = n_{\overline{xy}}/n$; $p_x = p_{xy} + p_{x\overline{y}} = 1 - p_{\overline{x}}$; $p_y = p_{xy} + p_{\overline{x}y} = 1 - p_{\overline{y}}$; $p_{\overline{x}} = p_{\overline{x}y} + p_{\overline{xy}} = 1 - p_x$; and $p_{\overline{y}} = p_{x\overline{y}} + p_{\overline{xy}} = 1 - p_y$.

Table 5.1 Absolute and
relative frequencies for any
association rule $X \rightarrow Y$
comprising an antecedent X
and a consequent Y in a
dataset Ω

	Y	$\overline{Y}$	Σ
(a) Absolute frequencies			
X	n_{xy}	$n_{x\overline{y}}$	n_x
$\overline{X}$	$n_{\overline{x}y}$	$n_{\overline{x}\,\overline{y}}$	$n_{\overline{x}}$
Σ	n_y	$n_{\overline{y}}$	$|\Omega|$
(b) Relative frequencies			
X	p_{xy}	$p_{x\overline{y}}$	p_x
$\overline{X}$	$p_{\overline{x}y}$	$p_{\overline{x}\,\overline{y}}$	$p_{\overline{x}}$
Σ	p_y	$p_{\overline{y}}$	1

5.2.1 Quality Measures

As it was previously defined, when mining association rules in a database comprising k items a total of $3^k - 2^{k+1} + 1$ solutions can be produced, or $(2^k - 1) \times t$ solutions if the consequent is a target variable including t distinct values. In such a set, a large percentage of solutions may be uninteresting and useless and a series of quality measures (see Table 5.2) to quantify the importance of each association rule is therefore required [45].

Support [2] is the most well-known quality measure in association rule mining. It was formally defined as the relative frequency of occurrence p_{xy} of an association rule $X \rightarrow Y$. This quality measure is symmetric, an important feature of any good metric. In other words, $Support(X \rightarrow Y) = Support(Y \rightarrow X)$ so support quality measure does not quantify any implication between the antecedent X and the consequent Y. Additionally, the minimum and maximum values for this quality measure are $p_{xy} = 0$ and $p_{xy} = 1$, respectively. In situations where $p_{xy} = 0$ or $p_{xy} = 1$, then the rule is useless since it does not provide any new unexpected knowledge about the data properties. In a recent study about quality measures in association rule mining [45], it was demonstrated that $p_{xy} = 0$ if and only if $p_x = 0$ or $p_y = 0$, or when the antecedent X and the consequent Y satisfy up to 50% of the data records within the dataset, i.e. $p_x \leq 0.5$ and $p_y \leq 0.5$, but they do not satisfy any record in common. Finally, it is remarkable to note that its value is always greater than 0 when $p_x + p_y > 1$, and its maximum value is always equal to the minimum value among p_x and p_y, i.e. $p_{xy} \leq Min\{p_x, p_y\}$.

Coverage [41] quality measure is formally defined as p_x, determining the percentage of data records where the antecedent X appears regardless the presence of Y. On the contrary, Prevalence [41] quality measure is defined as the percentage of data records where the consequent Y appears, i.e. p_y, regardless whether X also appears or not. In both metrics, i.e. Coverage and Prevalence, the feasible range of values is [0, 1].

Together with the Support quality measure, Confidence [48] is a really important metric within the associating rule mining field. Confidence identifies the percentage of data records that satisfy both X and Y at time once it is known that X was already satisfied. In a formal way, this quality measure can be expressed as

Table 5.2 Quality measures and range of feasible values

Quality measure	Equation	Feasible values
Support	p_{xy}	$[0, 1]$
Coverage	p_x	$[0, 1]$
Prevalence	p_y	$[0, 1]$
Confidence	$\dfrac{p_{xy}}{p_x}$	$[0, 1]$
Lift	$\dfrac{p_{xy}}{p_x \times p_y}$	$[0, n]$
Cosine	$\dfrac{p_{xy}}{\sqrt{p_x \times p_y}}$	$[0, 1]$
Leverage	$p_{xy} - (p_x \times p_y)$	$[-0.25, 0.25]$
Conviction	$\dfrac{p_x \times p_{\bar{y}}}{p_{x\bar{y}}}$	$[\dfrac{1}{n}, \dfrac{n}{4}]$
Centered confidence	$\dfrac{p_{xy}}{p_x} - p_y$	$[-1, 1-\dfrac{1}{n}]$
Certainty factor	$\begin{cases} \dfrac{\dfrac{p_{xy}}{p_x} - p_y}{p_{\bar{y}}} & if\ (\dfrac{p_{xy}}{p_x} - p_y) \geq 0 \\ \dfrac{\dfrac{p_{xy}}{p_x} - p_y}{p_y} & Otherwise \end{cases}$	$[-1, 1]$
Recall	$\dfrac{p_{xy}}{p_y}$	$[0, 1]$
Laplace	$\dfrac{p_{xy} \times n + 1}{p_x \times n + 2}$	$[\dfrac{1}{n_x + 2}, \dfrac{n_x + 1}{n_x + 2}]$
Pearson	$\dfrac{p_{xy} - (p_x \times p_y)}{\sqrt{p_x \times p_y \times p_{\bar{x}} \times p_{\bar{y}}}}$	$[-\dfrac{n_x \times n_y}{n_{\bar{x}} \times n_{\bar{y}}}, \dfrac{n_x \times n_{\bar{y}}}{n_{\bar{x}} \times n_y}]$
Information gain	$log(\dfrac{p_{xy}}{p_x \times p_y})$	$[log(\dfrac{1}{n}), log(n)]$
Sebag	$\dfrac{p_{xy}}{p_{x\bar{y}}}$	$[0, n - 1]$
Least contradiction	$\dfrac{p_{xy} - p_{x\bar{y}}}{p_y}$	$[-(n - 1), 1 - \dfrac{1}{n}]$
Example/counterexample rate	$1 - \dfrac{p_{x\bar{y}}}{p_y}$	$[2 - n, 1]$
Zhang	$\dfrac{p_{xy} - (p_x \times p_y)}{Max\{p_{xy} \times p_{\bar{x}},\ p_{xy} \times p_{x\bar{y}}\}}$	$[-1, 1]$
Netconf	$\dfrac{p_{xy} - (p_x \times p_y)}{p_x \times (1 - p_x)}$	$[-1, 1]$
Yule's Q	$\dfrac{p_{xy} \times p_{\bar{x}\bar{y}} - p_{x\bar{y}} \times p_{\bar{x}y}}{p_{xy} \times p_{\bar{x}\bar{y}} + p_{x\bar{y}} \times p_{\bar{x}y}}$	$[-1, 1]$

$Confidence(X \rightarrow Y) = p_{xy}/p_x$, or as an estimate of the conditional probability $P(Y|X)$. Unlike the Support quality measure, Confidence denotes an implication between X and Y since it is not a symmetric measure, i.e. $Confidence(X \rightarrow Y) \neq Confidence(Y \rightarrow X)$. Finally, it is important to highlight that Confidence takes values in the range $[0, 1]$.

The interest of an association rule has been studied by different authors and one of the most important metrics in this regard is the Lift quality measure [7]. Lift calculates the relationship between the Confidence of the rule and its Prevalence, i.e. $Lift(X \rightarrow Y) = Confidence(X \rightarrow Y)/p_y$ or $Lift(X \rightarrow Y) = p_{xy}/(p_x \times p_y)$. In other words, Lift calculates the ratio between the joint probability of two observing variables (antecedent X and consequent Y) with respect to their probabilities under the independence assumption. This metric is considered as a good correlation measure since it determines the degree of dependence between the antecedent and consequent of a rule. Here, Lift values lower than 1 determine a negative dependence; values greater than 1 denote a positive dependence; and values of 1 stands for independence. Focusing on the symmetry of this quality measure, its behaviour is similar to the Support metric since $Lift(X \rightarrow Y) = Lift(Y \rightarrow X)$ and, therefore, it does not quantify any implication between X and Y. Finally, as for the feasible range of values, Lift takes the value 0 when $p_{xy} = 0$ and $p_x \neq 0$ and/or $p_y \neq 0$, whereas its maximum value is obtained when $p_{xy} = p_x = p_y = 1/n$ (only a single data record is satisfied).

Tan et al. [41] proposed a quality metric, known as Cosine, that is derived from Lift. Cosine is formally defined as $Cosine(X \rightarrow Y) = \sqrt{Lift \times Support} = p_{xy}/\sqrt{p_x \times p_y}$ and its main difference with regard to Lift is the square root on the product of p_x and p_y. Similarly to Support and Lift, Cosine does not quantify any implication between the antecedent X and consequent Y since it is a symmetric measure, i.e. $Cosine(X \rightarrow Y) = Cosine(Y \rightarrow X)$. On the contrary, a major advantage of Cosine versus Lift is the range of feasible values, which is enclosed in the range $[0, 1]$ for Cosine whereas the upper bound is not predefined for Lift. Finally, it should be highlight that Cosine takes into account both the interestingness and the significance of an association rule since it contains two important quality measures in this sense, i.e. Support and Lift. Similarly, and based on the Lift quality measure, *Piatetsky-Shapiro* [36] propose the Leverage, which determines how different is the co-occurrence of the antecedent X and the consequent Y of a rule from independence [19]. In a forma way, Leverage is defined as $Leverage(X \rightarrow Y) = p_{xy} - (p_x \times p_y)$, taking values in the range $[-0.25, 0.25]$. Similarly to Lift, Leverage is symmetric, i.e. $Leverage(X \rightarrow Y) = Leverage(Y \rightarrow X)$, and it includes a value to denote statistical independence (zero value) between the antecedent and consequent.

The degree of implication of a rule has been studied by different researchers [7], giving rise to the Conviction quality measure. In a forma way, it is possible to express such a measure as $Conviction(X \rightarrow Y) = (p_x \times p_{\bar{y}})/p_{x\bar{y}}$. Additionally, the relative accuracy of a rule has been also considered in association rule mining, and an example is the Centered Confidence (CC), which is formally denoted as $CC(X \rightarrow Y) = Confidence(X \rightarrow Y) - p_y$.

All the aforementioned quality measures (Support, Coverage, Prevalence, Confidence, Lift, Cosine, Leverage, Conviction and CC) are considered as the most general or widely used measures in the association rule mining field. Nevertheless, it is also possible to find in the literature some quality measures not so generally used but of high importance in some contexts. An example is the Certainty Factor [5], which is defined as the CC measure but normalized into the interval $[-1, 1]$, i.e. $CF(X \rightarrow Y) = CC(X \rightarrow Y)/p_{\overline{y}}$ if $CC(X \rightarrow Y) \geq 0$), and $CF(X \rightarrow Y) = CC(X \rightarrow Y)/p_y$ if $CC(X \rightarrow Y) < 0$). Another example is Recall [11] which denotes the percentage of data records containing Y and, at the same time X. In a formal way, it is defined as $Recall(X \rightarrow Y) = p_{xy}/p_y$. Additional examples of quality measures have been recently described and analysed [45], including the following evaluation measures: Laplace measure is formally stated as $Laplace(X \rightarrow Y) = (n_{xy} + 1)/(n_x + 2)$; Pearson's correlation coefficient, which is defined as $(p_{xy} - (p_x \times p_y))/\sqrt{p_x \times p_y \times p_{\overline{x}} \times p_{\overline{y}}}$; Information Gain, denoted as $log(p_{xy}/(p_x \times p_y))$; Sebag quality measure that is defined as $p_{xy}/p_{x\overline{y}} = p_{xy}/(p_x - p_{xy})$; Least Contradiction, which is formulated as $(p_{xy} - p_{x\overline{y}})/p_y$; Example/Counterexample rate, defined as $1 - (p_{x\overline{y}}/p_y)$; Zhang et al. [48] proposed to quantify $(p_{xy} - (p_x \times p_y))/Max\{p_{xy} \times p_{\overline{x}}, p_{xy} \times p_{x\overline{y}}\}$; NetConf that is formally represented as $(p_{xy} - (p_x \times p_y))/(p_x \times (1 - p_x))$ to estimate the strength of an association rule; Yule'sQ defined as $(p_{xy} \times p_{\overline{xy}} - p_{x\overline{y}} \times p_{\overline{x}y})/(p_{xy} \times p_{\overline{xy}} + p_{x\overline{y}} \times p_{\overline{x}y})$.

5.2.2 Class Association Rules for Descriptive Analysis

CARs were formally defined, in a previous section, as specific cases of general association rules having a single item in the consequent. CARs are generally used for descriptive tasks, so given a set of records representing objects, customers, etc., and a property of interest (target variable), the aim of mining CARs is to find a set of items that properly describe the target variable. A real example of the discovery of CARs was recently proposed [28], aiming at describing the students' behaviour and how it is related to the final grade (target variable).

A major feature of CARs is that the simple fact of considering a single item or target variable in the consequent may help in improving the understanding of the provided knowledge [45]. In fact, this type of association rules are more frequently used than those including a large number of items in the consequent (general association rules). In general, results selected as of high interest in the association rule mining field tend to include a single item in the consequent, enabling a higher reliability since the support of the antecedent tend to be so similar to the one of the whole rule [26]. Some of the most well-known algorithms in association rule mining (Apriori [2] and FP-Growth [13]) can be used for mining CARs. Additionally, a wide number of approaches have been specifically proposed for this task [24, 27, 31, 32].

5.2.3 Class Association Rules for Predictive Analysis

In 1998, *Liu et al.* [22] connected association rule mining and classification rule mining to form the field known as associative classification (AC). For any predictive purpose, accurate and interpretable classifiers are desired by considering a pre-determined target variable, i.e. the class to be predicted. Here, thanks to CARs and by considering the class as the consequent of the rule, it is possible to take advantage of reliable and interesting association rules to form accurate and very interpretable classifiers. In order to obtain this kind of classifiers by using an AC algorithm, the complete set of CARs should be first discovered from the training data set and a subset of such rules is then selected to form the classifier. The selection of such a subset can be accomplished in many ways, and first approaches (CBA [22] or CMAR [21]) were based on selecting rules by using a database coverage heuristic [22]. Here, the complete set of CARs is evaluated on the training dataset and those rules that cover a certain number of training data records are considered. Recent AC techniques [43, 44] have been proposed to discover frequent itemsets (according to some minimum predefined support value), extract/rank CARs, and prune redundant CARs or those that lead to incorrect classification. Almost all the existing proposals so far are based on an exhaustive search methodology.

Although all the existing proposals work really great at the beginning, they present a series of drawbacks: exhaustive algorithms are not able to run in continuous features neither in large datasets (caused by the exponential complexity with regard to the number of items [30]). Besides, these proposals require two different steps for mining association rules, one for extracting frequent patterns and another for discovering reliable CARs among the previously mined patterns. Thus, to overcome the aforementioned problems, some researchers [3] have focused on the application of evolutionary algorithms for obtaining this kind of classifiers in few steps as well as in numerical domains.

5.2.4 Related Tasks

Supervised descriptive pattern mining includes a series of tasks aiming at searching for interesting data descriptions with respect to a property or variable of interest. In this book, contrast set mining [9] (see Chap. 2), emerging pattern mining [10] (see Chap. 3), and subgroup discovery [14] (see Chap. 4) have been already described as important descriptive tasks where a variable of interest is taken into account. In a previous analysis (see Sect. 4.2.2, Chap. 4) it was demonstrated that the definitions of contrast set mining, emerging pattern mining and subgroup discovery tasks were compatible.

Analysing the definition of CARs (discovery of sets of items that describe a property of interest) and comparing it with the definitions of the three aforemen-tioned techniques (contrast set mining, emerging pattern mining and subgroup discovery) some differences appear and they are mainly based on the terminology and used quality measures. At this point, it is noticeable that the mining of CARs and the discovery of subgroups really similar tasks. In fact, first algorithms for subgroup discovery were based on the extraction of all existing CARs in data as it was described in Sect. 4.3.2, Chap. 4. The first main difference lies in the quality measures defined for both tasks. Here, whereas subgroup discovery looks for unusual statistical distributions, the mining of CARs aims at seeking general and reliable rules that denote a co-occurrence between the antecedent and the target variable. The second major difference is related to the resulting sets, which are completely different for both tasks. Subgroup discovery aims at discovering a reduced set of items that properly describe the target variable and, for that, the unusual statistical distribution is quantified. On the contrary, the task of mining CARs aims at finding any rule that satisfies some quality thresholds, mainly based on Support and Confidence, with the only restriction that the target variable should be included in the consequent. Here, it is also important to note that the resulting set of any Subgroup Discovery algorithm may also be considered as a small set of CARs. On the contrary, the resulting set of any algorithm for mining CARs cannot be used in Subgroup Discovery. Instead, this resulting set needs to be analysed and reduced in a posterior step, just taking those solutions that denote an unusual statistical distribution.

5.3   Algorithms for Class Association Rules

The extraction of CARs have been generally used in two different tasks, that is, descriptive and predictive. In the following subsections, the most important approaches for these two tasks are described and analysed.

5.3.1   Algorithms for Descriptive Analysis

The task of mining frequent and reliable association rules is one of the most well-known and intensively researched data mining tasks [12]. The first algorithm for mining association rules is known as Apriori [2]. This algorithm is based on a level-wise paradigm in which all the frequent patterns of length $k + 1$ are generated by using all the frequent patterns of length k. Two (or more) patterns of length k can be joined to form a new $k + 1$ pattern if and only if they have, at least, $k - 1$ items in common. This algorithm can be easily modified (see Algorithm 5.1) to obtain CARs, where the only difference lies in the fact that the target variable is not considered as an item, but it is then considered to compute the frequency f of each pattern. Thus, any discovered itemset includes the a value of the target variable and, therefore, the algorithm should be run as many times as distinct values the target variable includes. As an example, let us consider a sample dataset (see Table 5.3) based on ten different

---

Algorithm 5.1 Pseudo-code of the Apriori algorithm for mining CARs

---

Require: $\Omega, t, mSupport$ ▷ dataset Ω, target value t and minimum support threshold
Ensure: P
1: $L_1 \leftarrow \{\forall i \in \Omega : f(i \wedge t) \geq mSupport\}$ ▷ frequent singletons by considering the target
   variable
2: $C \leftarrow \emptyset$
3: **for** $(k = 1; L_k \neq \emptyset; k++)$ **do**
4:    $C \leftarrow$ candidates patterns generated from L_k
5:    $L_{k+1} \leftarrow \{\forall c \in C : f(c \wedge t) \geq mSupport\}$ ▷ frequent patterns of size $k+1$ by
   considering the target variable
6: **end for**
7: **return** $P \leftarrow \cup_k L_k$

Table 5.3 Sample market
basket dataset

Items
{Bread}
{Bread, Beer}
{Bread, Beer, Diaper}
{Bread, Butter}
{Bread, Beer, Butter, Diaper}
{Beer}
{Beer, Diaper}
{Bread, Butter}
{Bread, Beer, Diaper}
{Bread}

records containing items (products) purchased by customers. In this example, the target variable is denoted as the item *Beer*. Thus, Apriori considers three singletons (*Bread*, *Butter* and *Diaper*) so a total of $2^3 - 1 = 7$ solutions can be found (each one should be considered together with the target variable). The Hasse diagram shown in Fig. 5.1 represents all the solutions, where the number of data records in which each pattern appears is illustrated into brackets and this number is dependent on the target variable. According to the aforementioned Hasse diagram, any pattern above the dashed line is a frequent pattern, whereas those below the dashed line are infrequent patterns.

Another well-known method for mining frequent itemsets is the FP-Growth algorithm [13]. Even when this algorithm was also proposed for mining frequent patterns and general association rules, it can be modified (see Algorithm 5.2) in order to extract patterns that are related to a target variable. FP-Growth is much more efficient than Apriori since it works on a compressed data representation in a tree form, and reduces the number of scans to be performed on data. The algorithm works by calculating the frequencies of each single item in data, which are sorted in descending order and saved into a list named *F-List*. Since this version is related to a target variable, the frequencies are calculated by considering each single item together with the target (see lines 2 and 3, Algorithm 5.2). In order to construct the tree structure, FP-Growth begins with the empty node *null* and the database

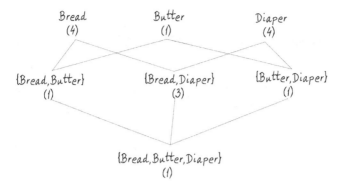

Fig. 5.1 Hasse diagram of a sample market basket dataset including ten different customers and considering *Beer* as target variable

---

Algorithm 5.2 Pseudo-code of the FP-Growth algorithm for mining CARs

---

Require: $\Omega, t, mSupport$ ▷ dataset Ω, target value t and minimum support threshold
Ensure: P
1: $Tree \leftarrow \emptyset$
2: $F\text{-}List \leftarrow \{\forall i \in \Omega : f(i \wedge t) \geq mSupport\}$ ▷ frequent singletons with the target variable
3: Sort $F\text{-}List$ in descending order of the frequency
4: Define the root of the tree data representation $Tree$
5: **for all** $r \in \Omega$ **do**   ▷ for each data record in Ω
6:      Make r ordered according to $F\text{-}List$
7:      Include the ordered record r in $Tree$ and considering t
8: **end for**
9: **for all** $i \in \Omega$ **do**   ▷ for each item in Ω
10:      $P \leftarrow$ Construct any pattern by taking i and traversing the $Tree$ structure
11: **end for**
12: **return** P

---

is scanned for each data record. Each of the analysed data records are ordered according to the *F-List* and, then every item is added to the tree. If a data record shares the items of any branch of the tree, then the inserted record will be in the same path from the root to the common prefix. Otherwise, new nodes are inserted in the tree by creating new branches, with support count initialized to one. This process ends when all data records have been inserted. Finally, the tree structure is analysed to obtain any frequent CAR since each pattern is related to a target value, so the algorithm should be run as many times as distinct values for the target variable.

Considering the sample market basket dataset (see Table 5.3), the FP-Growth algorithm analyses each data record in order to construct the frequent pattern tree structure (see Fig. 5.2). Each data record should include the target variable, i.e. *Beer* to be taken into account. The first, fourth, sixth, eighth and tenth data records are not considered since they do not include *Beer*. First, this algorithm ranks the singletons according to their frequencies to form a *F-List*, resulting as *Bread*, *Diaper* and *Butter*. Then, for each data record, items are analyzed and ordered according to the aforementioned *F-List*. Thus, for example, the second data record that includes

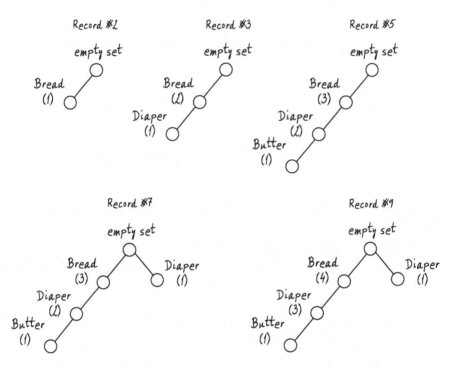

Fig. 5.2 Construction of the frequent pattern tree structure by using the sample market basket dataset shown in Table 5.3 and considering *Beer* as target variable

Bread and *Beer* (the target variable) is added to the tree structure. The third data record is then analyzed, which includes the items *Bread* and *Diaper* (together with the target variable *Beer*), and added to the tree structure. The process is repeated for each data record in Ω. Once the tree has been constructed, no further passes over the dataset are necessary and any CAR can be directly obtained by exploring the tree from the bottom-up. As a matter of clarification, the frequency of the pattern $P = \{Diaper, Bread\}$ is $f(P) = 2$ according to the tree structure (see Fig. 5.2), so there exist a CAR of the form $Diaper \wedge Bread \rightarrow Beer$ that is satisfied two times in data.

Some additional research studies [40] were focused on optimizing the Confidence quality measure (see Sect. 5.2.1). Given an association rule of the form $X \rightarrow Y$, Confidence [48] or reliability of the rule was defined as the percentage of data records that satisfy both X and Y at time once it is known that X was already satisfied. In a formal way, it is expressed as $Confidence(X \rightarrow Y) = p_{xy}/p_x$. Thus, the optimization of this evaluation measure entails that both p_{xy} and p_x should be almost the same, taking into account that $p_{xy} \geq p_x$. At this point and due to CARs only include a single item in the consequent Y, they tend to be really reliable rules and, therefore, they are generally used when the optimization of the Confidence is a dare [40].

Pattern mining [1] and the discovery of relationships between patterns [48] have been addressed by evolutionary computation [45], reducing the computational complexity and enabling continuous features to be mined without any discretization step. One of the first algorithms for representing continuous patterns by means of evolutionary computation and, more specifically, by means of genetic algorithms (GAs), was named GENAR (GENetic Association Rules) [32]. Here, authors represented each solution as a string of values where each two consecutive values were used to indicate the lower and upper bounds of a continuous interval. Thus, each individual (solution) is represented as a string of length equal to $2 \times k$, considering k as the number of distinct items or attributes in data, and the i-th pair of values is used to represent the i-th attribute in data. According to the authors, the last attribute is defined as the consequent of the rule, whereas the antecedent is defined by the remain attributes. Hence, GENAR was originally defined to mine only CARs where the target variable is defined in a continuous domain (see Fig. 5.3).

GENAR (see Algorithm 5.3) was designed to mine the best rule found in an evolutionary process (see lines 6 to 12). Thus, in this iterative algorithm, the evolutionary process is completed as many times as desired number of rules. GENAR ranks solutions according to the Support or frequency on the basis of those

Fig. 5.3 Example of an individual in GENAR to represent a set of continuous items or attributes

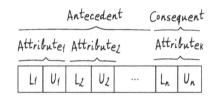

Algorithm 5.3 Pseudo-code of the GENAR algorithm

Require: $\Omega, mRules, mGen, pSize$ ▷ dataset Ω, rules to be discovered, maximum number of generations and population size
Ensure: *Rules*
1: $nRules \leftarrow 0$
2: **while** $nRules < mRules$ **do**
3:     $nGen \leftarrow 0$
4:     $offspring \leftarrow \emptyset$
5:     $population \leftarrow$ Generate $pSize$ random solutions
6:     Evaluate each solution within $population$ by considering Ω
7:     **while** $nGen < mGen$ **do**
8:        $parents \leftarrow$ select parents from $population$
9:        $offspring \leftarrow$ apply genetic operators to $parents$
10:       Evaluate each solution within $population$ by considering Ω
11:       $population \leftarrow$ take the best $pSize$ individuals from $\{population, offspring\}$
12:       $nGen \leftarrow nGen + 1$
13:     **end while**
14:     $Rules \leftarrow$ take the best rule in $population$
15:     $nRules \leftarrow nRules + 1$
16: **end while**
17: **return** *Rules*

records that were not satisfied by any other solution yet. Thus, more general CARs
are firstly obtained and keep in the set *Rules*.

Finally, it is important to remark that a major feature, and an important drawback
at the same time, of GENAR is that it works on a predefined pattern size, that is,
all the extracted solutions include the same set of items since the aim is to optimize
the bounds of the set of numerical attributes. Thus, in situations where the aim is
the looking for patterns with different sizes, the dataset needs to be modified just to
include the desired set of items, and individuals will contain only the given items.
Similarly, the dataset needs to be modified if the target variable was not originally
placed in the last attribute. To solve the problem of mining continuous patterns
that comprise a variable set of items, some authors have proposed the use of an
individual encoding that uses a variable-length string of values [33]. Unlike to the
previous individual representation (GENAR representation, see Fig. 5.3), a 3-tuple
of values is required now for each attribute so the first value is used to the item to
be represented. Thus, the sample 3-tuple (2, 0.50, 0.65) represents the 2nd attribute
defined in the range [0.50, 0.65].

Despite the fact that GAs have achieved a really good performance in the pattern
mining field, the way in which their solutions are encoded usually hampers the
mining process since the search space is not reduced [45]. The only restriction
considered by most GAs in this field is related to the size of the patterns to be
extracted. Considering the restriction of the search space as a baseline, the use
of grammars to include syntax constraints has been of great interest in recent
years [34]. In pattern mining, grammars have been considered as useful forms
of introducing subjective knowledge into the mining process, enabling patterns of
interest to be obtained by a variety of users with different goals.

The first use of grammars for mining patterns of interest and, specially, CARs,
was proposed by *Luna et al.* [24]. In this algorithm, known as G3PARM (Grammar-
Guided Genetic Programming for Association Rule Mining), authors defined a
grammar (see Fig. 5.4) where solutions comprised a set of items in the antecedent
of the rule, whereas the consequent is formed only by a single item. This grammar
enables both positive and negative patterns to be obtained. G3PARM encodes each
individual as a sentence of the language $(L(G) = \{Comparison\,(\wedge Comparison)^n$

$G = (\Sigma_N, \Sigma_T, P, S)$ with:

$\quad S \;\; = $ Rule
$\quad \Sigma_N = \{$Rule, Antecedent, Consequent, Comparison, Comparator, Attribute$\}$
$\quad \Sigma_T = \{'\wedge', '=', '\neq', 'name', 'value'\}$
$\quad P \;\; = \{$Rule = Antecedent, Consequent ;
$\qquad\qquad$ Antecedent = Comparison | '$\wedge$', Comparison, Antecedent ;
$\qquad\qquad$ Consequent = Comparison ;
$\qquad\qquad$ Comparison = Comparator, Attribute ;
$\qquad\qquad$ Comparator = '=' | '$\neq$' ;
$\qquad\qquad$ Attribute = '*name*', '*value*' ;$\}$

Fig. 5.4 Context-free grammar defined for the G3PARM algorithm

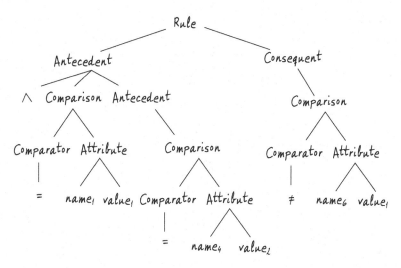

Fig. 5.5 Sample individual encoded by G3PARM by using the CFG shown in Fig. 5.4

$\rightarrow$ *Comparison*, $n \geq 0$}) generated by the grammar G (see Fig. 5.4), considering a maximum number of derivations to avoid extremely large trees. To obtain individuals (solutions to the problem), a set of derivation steps is carried out by applying the production rules declared in the set P and starting from the start symbol *Rule*. It should be noted that each terminal symbol randomly chooses the names (*name*) and values (*value*) from the dataset metadata. Thus, each individual in G3PARM is represented by a syntax-tree structure according to the defined grammar, and considering a maximum depth to avoid infinite derivations. Figure 5.5 illustrates a sample individual that includes either positive and negative items and which represents the rule $name_1 = value_1 \wedge name_4 = value_2 \rightarrow name_6 \neq value_1$.

G3PARM follows a traditional evolutionary schema (see Algorithm 5.4) including an elite population to keep the best individuals obtained along the evolutionary process. In each iteration of the evolutionary process, and for the sake of producing new individuals, G3PARM uses two of the most well-known genetic operators (see line 9, Algorithm 5.4) used in grammar-guided genetic programming. The crossover operator creates new solutions by exchanging compatible subtrees from pairs of individuals that act as parents. As for the mutation operator, a subtree of an individual is randomly chosen, producing a completely new subtree through the application of production rules to the node. Each new individual that is produced either at the beginning of the evolutionary process (see lines 5 and 6, Algorithm 5.4) or after the application of genetic operators (see lines 9 and 10, Algorithm 5.4) is evaluated by a fitness function that is based on the Support quality measure. One of the most important steps in G3PARM is the updating of the elite population *eliteP*. Here, the best solutions according to the Support measure (the fitness function) are kept into *eliteP* if they satisfy a minimum Confidence value *mConf* (see line

---

Algorithm 5.4 Pseudo-code of the G3PARM algorithm

Require: $\Omega, G, mRules, mGen, pSize, mConf$ $\triangleright$ dataset Ω, defined
    context-free grammar G, rules to be discovered, maximum number of generations, population
    size $pSize$ and minimum Confidence threshold value $mConf$
Ensure: *Rules*
  1: $P \leftarrow \emptyset$ $\triangleright$ population of individuals
  2: $eliteP \leftarrow \emptyset$ $\triangleright$ elite population that will include the best $mRules$ individuals
  3: $offspring \leftarrow \emptyset$
  4: $nGen \leftarrow 0$ $\triangleright$ current generation
  5: $P \leftarrow$ Generate $pSize$ random solutions by considering the grammar G $\triangleright$ see Figure 5.4
  6: Evaluate each solution within P by considering Ω
  7: **while** $nGen < mGen$ **do**
  8:      $parents \leftarrow$ select parents from P
  9:      $offspring \leftarrow$ apply genetic operators to $parents$
10:      Evaluate each solution within P by considering Ω
11:      $P \leftarrow$ take the best $pSize$ individuals from $\{P, offspring, eliteP\}$
12:      $eliteP \leftarrow$ the best $mRules$ individuals from $\{P\}$ that satisfy $mConf$
13:      $nGen + +$
14: **end while**
15: $Rules \leftarrow eliteP$
16: **return** *Rules*

---

12). Thus, the aim was to extract highly frequent patterns and reliable CARs from that patterns in just a single step, that is, rules are not produced from the set of frequent patterns in a second stage. Once all the generations have been performed, G3PARM returns the set *Rules* that comprises the set of best rules discovered along the evolutionary process (no more than $mRules$ can be returned).

The mining of CARs have been considered as a mechanism for extracting infrequent or rare patterns. Communication failure detection [38], analysis of interesting rare patterns in telecommunication networks [20], recognition of patients who suffer a particular disease that does not often occur [35], or credit card fraud detection [39] are some of the applications where it is interesting to use associations among patterns that do not frequently occur in data. The mining of rare association rules have been analysed from two perspectives [18]: (1) perfectly rare association rules (PRAR) are those that satisfy a minimum Confidence threshold value and their Support is lower than a maximum predefined value. Additionally, the Support value for each of the items included in the rule should be lower than the maximum predefined value. (2) imperfectly rare association rules (IRAR) are quite similar to perfectly association rules. The main difference lies in the fact that at least one of the items within the rule should have a Support value that is greater than the maximum predefined value.

In 2014, *Luna et al.* [27] proposed any interesting algorithm for mining not only PRAR but also IRAR. This approach, known as Rare-G3PARM, is a modified version of the previously described G3PARM algorithm [25]. In fact, the context-free grammar used by Rare-G3PARM to encode the solutions is similar to the one

$G = (\Sigma_N, \Sigma_T, P, S)$ with:

S = Rule
Σ_N = {Rule, Antecedent, Consequent, Comparison, Comparator_Discrete, Comparator_Continuous, Attribute_Discrete, Attribute_Continuous}
Σ_T = {'$\wedge$', '$=$', '$\neq$', '$<$', '$\leq$', '$>$', '$\geq$', '*name*', '*value*'}
P = {Rule = Antecedent, Consequent ;
          Antecedent = Comparison | '$\wedge$', Comparison, Antecedent ;
          Consequent = Comparison ;
          Comparison = Discrete, Attribute | Continuous, Attribute ;
          Discrete = '$\neq$' | '$=$' ;
          Continuous = '$<$' | '$\leq$' | '$>$' | '$\geq$' ;
          Attribute = '*name*', '*value*' ;}

Fig. 5.6 Context-free grammar defined for the Rare-G3PARM algorithm

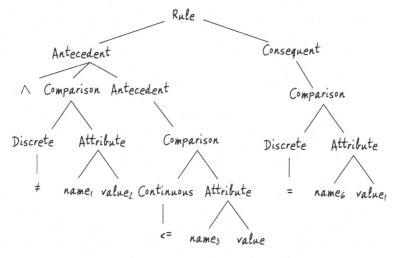

Fig. 5.7 Sample individual encoded by Rare-G3PARM by using the CFG shown in Fig. 5.6

of G3PARM but considering both continuous and discrete items (see Fig. 5.6). Figure 5.7 illustrates a sample individual that includes positive, negative and continuous items and which represents the rule $name_1 = value_1 \wedge name_4 = value_2 \rightarrow name_6 \neq value_1$.

Rare-G3PARM (see Algorithm 5.5) determines that an individual is considered to be included in the elite population if and only if it satisfies some minimum quality thresholds for the Confidence and Lift metrics (see line 12, Algorithm 5.5). Additionally, solutions are evaluated according to four different fitness functions that are based on Support. The first fitness function (see Eq. (5.1)) provides a maximum value in the middle of a specific interval that ranges between a minimum and a maximum Support value. This fitness function takes low values for Support values that are close to the boundaries. A zero value is assigned out of the interval.

Algorithm 5.5 Pseudo-code of the Rare-G3PARM algorithm

Require: $\Omega, G, mRules, mGen, pSize, mConf, mLift$ $\triangleright$ dataset Ω, defined context-free
   grammar G, rules to be discovered, maximum number of generations, population size $pSize$,
   minimum Confidence threshold value $mConf$ and minimum Lift threshold value $mLift$

Ensure: $eliteP$
1:   $P \leftarrow \emptyset$
2:   $eliteP \leftarrow \emptyset$
3:   $offspring \leftarrow \emptyset$
4:   $nGen \leftarrow 0$ $\triangleright$ current generation
5:   $P \leftarrow$ Generate $pSize$ random solutions by considering the grammar G $\triangleright$ see Figure 5.6
6:   Evaluate each solution within P by considering Ω
7:   **while** $nGen < mGen$ **do**
8:       $parents \leftarrow$ select parents from P
9:       $offspring \leftarrow$ apply genetic operators to $parents$
10:      Evaluate each solution within P by considering Ω
11:      $P \leftarrow$ take the best $pSize$ individuals from $\{P, offspring, eliteP\}$
12:      $eliteP \leftarrow$ the best $mRules$ individuals from $\{P\}$ that satisfy $mConf$ and $mLift$
13:      $nGen++$
14:   **end while**
15:   **return** $eliteP$

$$F_1(rule) = \begin{cases} \dfrac{Support(rule) - Min}{\dfrac{(Max + Min)}{2}} & \text{if } Min \leq Support(rule) \leq \dfrac{(Max + Min)}{2} \\[4mm] \dfrac{Max - Support(rule)}{\dfrac{(Max + Min)}{2}} & \text{if } \dfrac{(Max + Min)}{2} \leq Support(rule) \leq Max \\[4mm] 0 & \text{Otherwise} \end{cases}$$

$$(5.1)$$

The second fitness function (see Eq. (5.2)) is based on the fact that any solution having a Support value within a predefined interval is equally promising, being essential to establish an additional mechanism to properly differentiate among rules. Here, Confidence and Lift measures are of high interest, ranking the solutions based on these two measures. Finally, the other two fitness functions (see Eqs. (5.3) and (5.4)) were defined to minimize and maximize the support within a given Support interval, respectively. One of these fitness functions (see Eq. (5.3)) provides high values for those solutions having a Support value close to the lower boundary. The other fitness function (see Eq. (5.4)), on the contrary, provides high values for those solutions having a Support value close to the upper boundary.

$$F_2(rule) = \begin{cases} 1 & \text{if } Min \leq Support(rule) \leq Max \\ 0 & \text{Otherwise} \end{cases} \qquad (5.2)$$

$$F_3(rule) = \begin{cases} 1 + \dfrac{Min - Support(rule)}{Max - Min} & \text{if } Min \leq Support(rule) \leq Max \\[3mm] 0 & \text{Otherwise} \end{cases} \qquad (5.3)$$

$$F_4(rule) = \begin{cases} \dfrac{Support(rule) - Min}{Max - Min} & \text{if } Min \leq Support(rule) \leq Max \\ 0 & \text{Otherwise} \end{cases} \qquad (5.4)$$

In data analytics, it is of high interest the mining of context-sensitive features [6], that is, features whose meaning depends on the context in which they are defined. *Widmer* [46] described a context-sensitive feature as an adjustable filter for giving the right meaning of a concept in a specific context and, what is more important, a chunk of information may be meaningful or meaningless depending on this context. When dealing with descriptive analytics, the context may provide accurate, powerful and actionable insights to understand the relationships among patterns. CARs have been recently applied for dealing with contextual features [31], seeking associations of the form $Context \wedge X \rightarrow y$ such that the strength of its implication depends on some contextual information.

According to *Luna et al.* [31], traditional algorithms for mining associations between patterns are not aware of the context in which they are described even when many features are usually context-sensitive in real-world domains [6]. As a matter of clarification, let us consider the rule $diapers \rightarrow beer$ that is one of the most well-known associations defined in the market basket analysis, denoting that it is highly probable that someone buys beer and diapers at time. It would allow shop-keepers to exploit this specific relationship by moving their products closer together on the shelves. However, from the context-sensitive point of view, the meaning of the rule may vary since diapers are not only used for babies. Thus, the meaning of the rule for an elder caregiver is completely different to the meaning for a young father. An elder caregiver should not drink while looking after an elderly person, so shop-keepers should not exploit this relationship to increase the sells. All of this leads us to the conclusion that the analysis of the context is primary and, sometimes, required to provide an accurate insight that is able to produce the right meaning.

In a formal way, a contextual rule was formally defined [31] as an implication of the form $X_i \wedge X_n \rightarrow y$ such as the probability distribution of a concept y defined in the set Y of concepts is different than the obtained when the feature X_i comes into play, assuming that we know the values of other features X_n. Additionally, an irrelevant association rule $X_i \wedge X_n \rightarrow y$ was formally defined as those associations in which the probability distribution of a concept y defined in the set Y of concepts is not affected by the feature X_i, assuming that we know the values of other features X_n.

For the extraction of this type of associations, authors [31] proposed a model that is based on a grammar-guided genetic programming methodology. In this algorithm authors defined a grammar (see Fig. 5.8) where solutions comprised a set of items in the antecedent of the rule, whereas the consequent is formed only by a single item. This grammar enables both discrete and continuous features to be considered, and each individual is encoded as a sentence of the language $(L(G) = \{ Condition (\wedge Condition)^n \rightarrow Condition, n \geq 0\})$ generated by the grammar G (see Fig. 5.8), considering a maximum number of derivations to avoid extremely large trees.

$G = (\Sigma_N, \Sigma_T, P, S)$ with:

$S \quad = R$
$\Sigma_N = \{R, \text{Conditions, Condition, Discrete, Continuous} \}$
$\Sigma_T = \{ `\wedge', \text{`Attribute'}, `=', \text{`IN'}, \text{`value'}, \text{`Min\_value'}, \text{`Max\_value'} \}$
$P \quad = \{\text{Rule = Conditions, Condition} ;$
$\quad\quad \text{Conditions = Condition, } `\wedge', \text{ Condition} ;$
$\quad\quad \text{Condition = Discrete } | \text{ Continuous} ;$
$\quad\quad \text{Discrete = `Attribute', } `=', \text{ `value'} ;$
$\quad\quad \text{Continuous = `Attribute', `IN', `Min\_value', `Max\_value'} ;\}$

Fig. 5.8 Context-free grammar defined for mining contextual association rules

---

Algorithm 5.6 Pseudo-code of the algorithm for mining contextual rules

---

Require: $\Omega, G, mRules, mGen, pSize$ $\triangleright$ dataset Ω, defined context-free grammar G, rules to be discovered, maximum number of generations and population size $pSize$
Ensure: $eliteP$
1: $P \leftarrow \emptyset$
2: $I \leftarrow \emptyset$ $\triangleright$ population including irrelevant features
3: $eliteP \leftarrow \emptyset$
4: $offspring \leftarrow \emptyset$
5: $nGen \leftarrow 0$ $\triangleright$ current generation
6: **while** number of individuals in $P < pSize$ **do**
7: $\quad$ $s \leftarrow$ generate a random solution by considering the grammar G $\triangleright$ see Figure 5.8
8: $\quad$ Evaluate the solution s by considering Ω
9: $\quad$ **if** $f(s) > 0$ **then**
10: $\quad\quad$ $P \leftarrow$ add s to the population
11: $\quad$ **end if**
12: $\quad$ **if** $f(s) = -1$ **then**
13: $\quad\quad$ $I \leftarrow$ add s to the population of irrelevant features
14: $\quad$ **end if**
15: **end while**
16: **while** $nGen < mGen$ **do**
17: $\quad$ $parents \leftarrow$ select parents from P
18: $\quad$ $offspring \leftarrow$ apply genetic operators to $parents$
19: $\quad$ Evaluate each solution within P by considering Ω
20: $\quad$ **if** $\exists s \in P : f(s) = -1$ **then**
21: $\quad\quad$ $I \leftarrow$ add s to the population of irrelevant features
22: $\quad$ **end if**
23: $\quad$ $P \leftarrow$ take the best $pSize$ individuals from $\{P, offspring, eliteP\}$
24: $\quad$ $eliteP \leftarrow$ the best $mRules$ individuals from $\{P\}$
25: $\quad$ $nGen + +$
26: **end while**
27: **return** $eliteP$

The proposed algorithm (see Algorithm 5.6) starts by encoding individuals that represent CARs through the use of a context-free grammar defined in Fig. 5.8, and each individual is evaluated according to the function $lift(X \rightarrow y) - \sum_{k=1}^{n} lift(X_k \rightarrow y)$. In this first stage of the model, the aim is to produce feasible solutions by generating a desired number of individuals having a fitness value greater than 0 (see lines 6 to 15, Algorithm 5.6). If the fitness value for a generated individual is 0, then a new individual is produced till the desired number is reached. On the contrary, if a specific individual has a fitness function value equal to -1, then this individual includes an irrelevant feature and it is kept as a rule of interest for the user by keeping it in a set of irrelevant individuals or solutions. The ultimate aim of this first stage is to improve the evolutionary process since it starts with a population of diverse but feasible individuals.

This algorithm for mining contextual CARs follows a traditional evolutionary schema (see Algorithm 5.6) including an elite population to keep the best individuals obtained along the evolutionary process. In each generation of this evolutionary process (see lines 16 to 26, Algorithm 5.6), new individuals are generated through the application of two genetic operators (standard crossover and mutation operators were considered). A major feature of this algorithm is it does not need a fixed probability value but it self-adapts the value based on the diversity of the population, checking whether the average fitness value of the population was increased or decreased during the previous generations. In such iterations where the average fitness value is being improved, then a depth search is required so the crossover probability value is increased and the mutation probability value is decreased. On the contrary, if the fitness value is not being improved, then a higher diversity is required so the crossover probability value is decreased and the mutation probability value is increased.

5.3.2 Algorithms for Predictive Analysis

CARs have been widely used for building accurate and interpretable classifiers [42]. The integration of two well-known data mining tasks (association rule mining and classification rule mining) was first proposed in 1998 by *Liu et al.* [22]. Authors proposed a special subset of association rules, i.e. CARs, whose right-hand-side are restricted to the classification class attribute. According to the authors [22], the resulting classifiers helps in solving a number of important existing problems with the classification systems. First, it may solve the understandability problem produced by standard classification systems, which produce a large number of rules difficult to understand since they tend to be domain independent. On the contrary, building a classifier through CARs (all the rules are generated) is reduced to a post-processing task by applying specific techniques that help the user identify understandable rules. Second, the looking for a small set of rules in classification systems results in many interesting and useful rules not being discovered (rules including a specific feature that is essential for the end user). Finally, traditional classification approaches may be too sensitive to small changes in the input data.

The first algorithm in this field was proposed by *Liu et al.* [22]. This approach, named CBA (Classification Based on Associations), consists of two main stages: a rule generator based on the Apriori algorithm [2], and a classifier builder. In the first phase, a modified version of Apriori (see Algorithm 5.7) is applied in which each set of items are evaluated with an associated target value (the class attribute). In this first

---

Algorithm 5.7 Pseudo-code of the rule generator procedure of the CBA algorithm

---

Require: $\Omega, t, mSupport, mConfidence$ ▷ dataset Ω, target value t, minimum support and
 confidence thresholds

Ensure: P

1: $F_1 \leftarrow \{\forall i \in \Omega : f(i \wedge t) \geq mSupport\}$ ▷ frequent singletons by considering the target
 variable
2: $P \leftarrow \{\forall i \in F_1 : f(i \wedge t)/f(i) \geq mConfidence\}$ ▷ reliable rules
3: $C \leftarrow \emptyset$
4: **for** $(k = 1; L_k \neq \emptyset; k ++)$ **do**
5:     $C \leftarrow$ candidates patterns generated from L_k
6:     $F_{k+1} = \{\forall c \in C : f(c \wedge t) \geq mSupport\}$ ▷ frequent patterns of size $k + 1$ by
 considering the target variable
7:     $P \leftarrow$ add the following set $\{\forall i \in F_{k+1} : f(i \wedge t)/f(i) \geq mConfidence\}$ ▷ reliable rules
8: **end for**
9: **return** P

Algorithm 5.8 Pseudo-code of the classifier builder procedure of the CBA algorithm

---

Require: Ω, P ▷ dataset Ω and set of frequent and reliable rules

Ensure: C

1: Sort rules from P according to Confidence, Support, generation order
2: **for all** $p \in P$ **do**
3:     $temp \leftarrow \emptyset$
4:     **for all** $r \in \Omega$ **do**
5:         **if** r satisfies the conditions of p **then**
6:             $temp \leftarrow$ add r and mark p if it correctly classifies r
7:         **end if**
8:     **end for**
9:     **if** p is marked **then**
10:         $C \leftarrow$ add p
11:         $\Omega \leftarrow \Omega \setminus temp$
12:         Select a default class for the current C
13:         Compute the total number of errors of C
14:     **end if**
15: **end for**
16: Find the first rule p in C with the lowest total number of errors and remove all the rules after
 p in C
17: Add the default class associated with p to end C
18: **return** C

---

stage, minimum support and confidence thresholds (see Sect. 5.2.1) are considered
so all the feasible CARs are both frequent and accurate. Finally, the second stage is
responsible for producing the best classifier out of the whole set of rules previously
obtained. Here, it is required to evaluate all the possible subsets of discovered rules
on the training data and check the right rule sequence that gives the least number
of errors. However, it is not possible to perform it in an exhaustive manner so a
heuristic is performed (see Algorithm 5.8). In this second procedure, rules from the
set P are first sorted (see line 1, Algorithm 5.8) according to a series of criteria:
$p_i \in P$ precedes $p_j \in P$ if the Confidence value of p_i is greater than that of p_j; if
the Confidence values are the same, then p_i precedes p_j if the Support value of p_i

is greater than that of p_j; if both values are the same (Support and Confidence), then the one that was first generated will precede the other one. In the classifier builder procedure, each rule is analyzed to find those data records that are covered by the rule, removing such data records and selecting as default class the majority class in the remaining data. Finally, the procedure ends by discarding those rules in C that do not improve the final accuracy of the resulting classifier (see lines 14 and 16, Algorithm 5.8).

Liu et al. [23] proposed an improved version of the CBA algorithm, known as CBA2. According to the authors, two main weaknesses of building classifiers through CARs are inherit from association rule mining. First, the use of a single minimum support threshold for the rule generation is not enough for unbalanced class distributions. To understand this problem, let us use a sample dataset having two classes c_1 and c_2 such as c_1 appears in 95% of the cases and c_2 in the remain 5%. Considering a minimum Support threshold value of 6% no rule of the class c_2 will be obtained. On the contrary, let us consider a minimum Support threshold value of 0.1%, which is too low for the class c_1. Hence, it may find many overfitting rules for such a class. Finally, the second main problem described by *Liu et al.* [23] is related to classification data often contains a huge number of rules and, sometimes, the rule generator procedure is unable to generate rules with many conditions that may be important for producing an accurate classification.

In order to solve the first of the two aforementioned problems, authors of CBA2 [23] proposed the use of multiple thresholds for the Support quality measure in the rule generation phase (see Algorithm 5.9). Thus, a different threshold is assigned to each class depending on its frequency within data and following the formula $minSup(c_i) = minSup \times frequency(c_i)$; $minSup$ denoting the general Support threshold value, and $frequency(c_i)$ the frequency of each class within data. As a matter of example, let us consider the two aforementioned classes c_1 (appears in 95% of the cases) and c_2 (appears in 5% of the cases), and the general Support threshold value of 0.4. According to CBA2, this threshold value

---

Algorithm 5.9 Pseudo-code of the rule generator procedure of the CBA2 algorithm

Require: $\Omega, t, mSupport, mConfidence$ ▷ dataset Ω, target value t, minimum support and confidence thresholds
Ensure: P
1: $F_1 \leftarrow \{\forall i \in \Omega : f(i \wedge t) \geq (mSupport \times f(t))\}$ ▷ frequent singletons by considering the target variable and the Support threshold associated to this target variable
2: $P \leftarrow \{\forall i \in F_1 : f(i \wedge t)/f(i) \geq mConfidence\}$ ▷ reliable rules
3: $C \leftarrow \emptyset$
4: **for** $(k = 1; L_k \neq \emptyset; k++)$ **do**
5:    $C \leftarrow$ candidates patterns generated from L_k
6:    $F_{k+1} = \{\forall c \in C : f(c \wedge t) \geq (mSupport \times f(t))\}$ ▷ frequent patterns of size $k+1$ by considering the target variable and the Support threshold associated to this target variable
7:    $P \leftarrow$ add the following set $\{\forall i \in F_{k+1} : f(i \wedge t)/f(i) \geq mConfidence\}$ ▷ reliable rules
8: **end for**
9: **return** P

Algorithm 5.10 Pseudo-code of the rule generator procedure of the CBA2 algorithm

---

Require: C, T, N ▷ classifiers based on CBA2, C4.5 and Naive-Bayes, respectively
Ensure: C
 1: **for all** e in Ω **do**   ▷ for each example in training
 2:      Find the first rule r_i in CBA2 that covers e
 3:      **if** r classifies e wrongly **then**
 4:          Increases $Error_i$ by one
 5:      **end if**
 6:      **if** C4.5 classifies e wrongly **then**
 7:          Increases $Error_{i,T}$ by one
 8:      **end if**
 9:      **if** Naive-Bayes classifies e wrongly **then**
10:          Increases $Error_{i,N}$ by one
11:      **end if**
12: **end for**
13: **for all** r_i in C **do**   ▷ for each rule within the classifier
14:      **if** $Error_i \leq Error_{i,T}$ and $Error_i \leq Error_{i,N}$ **then**
15:          Keep r_i in C
16:      **else**
17:          **if** $Error_{i,T} \leq Error_{i,N}$ **then**
18:              Use the condition of r_i with the prediction given by T
19:          **else**
20:              Use the condition of r_i with the prediction given by N
21:          **end if**
22:      **end if**
23: **end for**
24: **return** C

---

is transformed into $minSup(c_1) = 0.4 \times 0.95 = 0.38$ for the class c_1, and $minSup(c_2) = 0.4 \times 0.05 = 0.02$ for the class c_2. It should be noted that this problem does not affect the Confidence measure so its threshold is the same for any of the classes. Finally, CBA2 includes a procedure that improve the rule generation phase by enabling long rules (rules with many conditions) to be discovered, which is a problem inherent to highly correlated datasets due to their large search spaces (combinatorial explosion). Due to such long rules may be important for classifications (the final classifiers suffer), CBA2 includes a combination technique (see Algorithm 5.10) to reduce the effect of the problem by considering both decision tree [37] and Naive-Bayes [8] methods. Thus, the proposed combination approach is based on the competition of three different classifiers on different segments of the training data. The key idea is to use one classifier to segment the training data, and then choose the best classifier to classify each segment. For the set of training examples covered by a rule r_i in C, the classifier that has the lowest error on the set of examples is selected to replace r_i.

In 2001, *Pei et al.* [21] proposed the CMAR (Classification based on Multiple Association Rules) algorithm for associative classification. Instead of using Apriori for mining CARs, CMAR extends an efficient frequent pattern mining method like FP-Growth [13] for the rule generation phase. In this phase, CMAR scans the training dataset once and keeps the items that appear in data at least a minimum

Table 5.4 Sample dataset

Items	Class
$item_1, item_3$	A
$item_1, item_2, item_3$	B
$item_4$	A
$item_1, item_2, item_4$	C
$item_1, item_2, item_3, item_4$	C

Algorithm 5.11 Pseudo-code of the classifier builder procedure of the CMAR algorithm

---

Require: Ω, R, δ ▷ dataset Ω, set of frequent and reliable rules and coverage threshold δ
Ensure: C
1: Sort rules from R according to the rank in descending order
2: **for all** $e \in \Omega$ **do** ▷ for each data record (example e) in Ω
3:   Initialize to 0 the cover-counter for e
4: **end for**
5: **while** $R \neq \emptyset$ and $\Omega \neq \emptyset$ **do**
6:   **for all** $r \in R$ **do** ▷ for each rule r in rank descending order
7:     Find all data records $E \subseteq \Omega$ matching r
8:     Increase the cover-count of E by 1
9:     Remove those data records having a cover-counter greater than δ
10:   **end for**
11: **end whilereturn** C

---

number of times. All other items, which fail the Support threshold, cannot play any role in the CARs, and thus can be pruned. Then, all the frequent items are sorted in Support descending order in this phase, forming an F-list. Finally, the training dataset is scanned again to construct an FP-Tree, which is a prefix tree in which items are extracted and sorted according to the F-list. As a matter of example, let us consider a sample dataset (see Table 5.4) with a Support threshold of 0.4 in per unit basis. Here, the list of frequent items (sorted in descending order of the Support measure) is $\{item_1, item_2, item_3, item_4\}$. Taking the first data record (see Fig. 5.9), a new branch is inserted into the tree and the class label (and the frequency) is attached to the last node in the path. Those sets that share items will also share branches within the tree.

Once a set of rules are obtained by CMAR, the next step is building an accurate classifier (see Algorithm 5.11). Since the number of generated CARs can be huge, CMAR needs to prune rules to delete redundant and noisy information. In this regard, a ranking is considered were giving two rules R_1 and R_2, R_1 is said to have a higher rank than R_2 if and only if: (a) Confidence(R_1) > Confidence(R_2); (b) Confidence(R_1) = Confidence(R_2) but Support(R_1) > Support(R_2); and, finally, (c) Confidence(R_1) = Confidence(R_2), Support(R_1) = Support(R_2) but R_1 has fewer items than R_2 does.

A similar algorithm was proposed under the name L^3 (Live and Let Live) [4]. In this research work, authors argued that rules may be useful to cover new cases

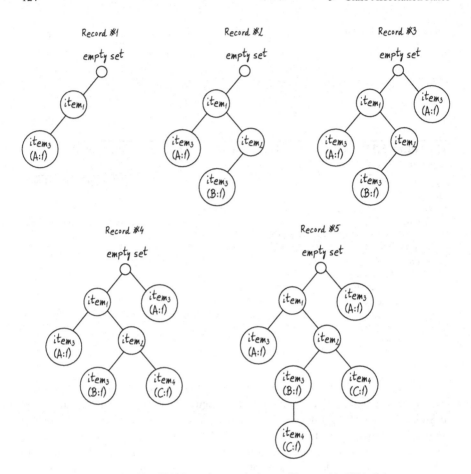

Fig. 5.9 FP-Tree obtained by CMAR on the sample dataset illustrated in Table 5.4

even if they have not bee used to cover training data and, therefore, should not be discarded. This algorithm includes a pruning strategy that only takes place when it is strictly necessary. The proposed pruning technique was designed to perform a reduced amount of pruning by eliminating only harmful rules, that is, those rules that only misclassify training data. During the process of building a classifier, L^3 adopts a two-steps methodology in which high-quality rules (rules used in the classification of the training data) are considered first, whereas unchecked rules (rules that were not unused during the training phase) are considered next to classify unlabeled data. In the classification builder procedure, and similarly to previous approaches, e.g. CMAR [21], rules are sorted according to Confidence and Support quality measures. However, and differently to previous works, when two rules have the same Confidence and Support values in L^3, then the rule with the highest number of items is desired since it covers the same training data but includes more constraints (it is considered more accurate).

In this pruning approach, discovered rules that were saved in the set R are considered one by one in sort order. For each rule $r \in R$, the set of records in Ω that are covered by r are selected and this set is not considered further. The rule $r \in R$ is pruned if it only performs wrong classifications. This process continues till all the rules in R are analyzed or there is no remain data records to be analyzed. It should be noted that, since rules are considered in sort order, a rule only matches training data not yet covered by any rule preceding it. Thus, it is possible that some rules do not match any record and these rules are included in a different set ($Level - II$ set) to be exploited in a posteriori step. To classify an unlabeled data d, those rules that covered at least one data record are considered in order and the first rule that matches d labels it. In such situations where no rule of this set matches d, then those rules saved in $Level - II$ are considered and the first rule that matches d labels it. If no rule matches d, then d is not labeled.

Finally, a well-known algorithm in this field was proposed by *Xin et al.* [47] under the name CPAR (Classification based on Predictive Association Rules). Methodology for this approach is quite similar to the one of the aforementioned algorithms, except for CPAR uses a greedy approach in the rule generation phase, which is much more efficient than generating all candidate rules. Then, similarly to most of the existing approaches in this field [43, 44], CPAR repeatedly searches for the current best rule and removes all the data records covered by the rule until there is no uncovered data record.

Associative classification has been also considered by means of fuzzy systems [17], enhancing the interpretability of the obtained classification rules and avoiding unnatural boundaries in the partitioning of continuous attributes. In this regard, authors in [3] proposed a Fuzzy Association Rule-based Classification method for High-Dimensional problems (FARC-HD) to obtain an accurate and compact fuzzy rule-based classifier with a low computational cost. FARC-HD (see Algorithm 5.12) comprises three different stages: (1) different fuzzy association rules are extracted, which should be as smaller (few number of items) as possible. Here, the minimum support for each class c_i is calculated as $minSup(c_i) = minSup \times frequency(c_i)$, where $minSup$ is the general support threshold value considered for the whole dataset; (2) Interesting rules are preselected by considering a subgroup discovery approach and considering a pattern weighting scheme [16]; (3) Finally, a selection of fuzzy association rules is performed to obtain an accurate classifier.

Algorithm 5.12 Pseudo-code of the FARC-HD algorithm

Require: $\Omega, q, minSupport, k$ ▷ dataset Ω, q predefined linguistic terms, minimum support
 threshold $minSupport$ and number of times k a record can be covered by different rules
Ensure: C
 1: Calculate the minimum support threshold for each class c_i within Ω as $minSup(c_i) = minSup \times frequency(c_i)$
 2: $R \leftarrow$ Obtain frequent rules with a maximum length (number of items)
 3: **for all** $c_i \in \Omega$ **do**   ▷ for each class (target attribute) c_i in Ω
 4:     Set the weight of the records containing c_i to 1
 5:     **for all** $r \in R$ **do**   ▷ for each rule r in R
 6:         Calculate the weighted relative accuracy for r ▷ This measure was described in
        Section 4.2.1, Chapter 4
 7:     **end for**
 8:     **while** there are more candidate rules or not all the records have been covered more than k
    times **do**
 9:         Select the best rule r and decrease the weights of the records covered by r
10:         $S \leftarrow r$
11:     **end while**
12: **end for**
13: Apply a CHC genetic algorithm to form an accurate classifier C from the set of rules S
14: **return** C

References

1. C.C. Aggarwal, J. Han, *Frequent Pattern Mining* (Springer International Publishing, Cham, 2014)
2. R. Agrawal, T. Imielinski, A.N. Swami, Mining association rules between sets of items in large databases, in *Proceedings of the 1993 ACM SIGMOD International Conference on Management of Data, SIGMOD Conference '93*, Washington, 1993, pp. 207–216
3. J. Alcalá-Fdez, R. Alcalá, F. Herrera, A fuzzy association rule-based classification model for high-dimensional problems with genetic rule selection and lateral tuning. IEEE Trans. Fuzzy Syst. **19**(5), 857–872 (2011)
4. E. Baralis, S. Chiusano, P. Garza, A lazy approach to associative classification. IEEE Trans. Knowl. Data Eng. **20**(2), 156–171 (2008)
5. F. Berzal, I. Blanco, D. Sánchez, M.A. Vila, Measuring the accuracy and interest of association rules: a new framework. Intell. Data Anal. **6**(3), 221–235 (2002)
6. P. Brézillon, Context in problem solving: a survey. Knowl. Eng. Rev. **14**(01), 47–80 (1999)
7. S. Brin, R. Motwani, J.D. Ullman, S. Tsur, Dynamic itemset counting and implication rules for market basket data, in *Proceedings of the 1997 ACM SIGMOD International Conference on Management of Data, SIGMOD '97*, Tucson, 1997 (ACM, New York, 1997), pp. 255–264
8. P.M. Domingos, M.J. Pazzani, On the optimality of the simple Bayesian classifier under zero-one loss. Mach. Learn. **29**(2–3), 103–130 (1997)
9. G. Dong, J. Bailey (eds.), *Contrast Data Mining: Concepts, Algorithms, and Applications* (CRC Press, Boca Raton, 2013)
10. A.M. García-Vico, C.J. Carmona, D. Martín, M. García-Borroto, M.J. del Jesus, An overview of emerging pattern mining in supervised descriptive rule discovery: taxonomy, empirical study, trends and prospects. WIREs Data Min. Knowl. Discov. **8**(1), e1231 (2017)
11. L. Geng, H.J. Hamilton, Interestingness measures for data mining: a survey. ACM Comput. Surv. **38**, Article 9 (2006)
12. J. Han, M. Kamber, *Data Mining: Concepts and Techniques* (Morgan Kaufmann, Amsterdam, 2000)

13. J. Han, J. Pei, Y. Yin, R. Mao, Mining frequent patterns without candidate generation: a frequent-pattern tree approach. Data Min. Knowl. Discov. **8**, 53–87 (2004)
14. F. Herrera, C.J. Carmona, P. González, M.J. del Jesus, An overview on subgroup discovery: foundations and applications. Knowl. Inf. Syst. **29**(3), 495–525 (2011)
15. S.J. Jung, C.S. Son, M.S. Kim, D.J. Kim, H.S. Park, Y.N. Kim, Association rules to identify complications of cerebral infarction in patients with atrial fibrillation. Healthc. Inform. Res. **19**(1), 25–32 (2013)
16. B. Kavšek, N. Lavrač, APRIORI-SD: adapting association rule learning to subgroup discovery. Appl. Artif. Intell. **20**(7), 543–583 (2006)
17. K. Kianmehr, M. Kaya, A.M. ElSheikh, J. Jida, R. Alhajj, Fuzzy association rule mining framework and its application to effective fuzzy associative classification. Wiley Interdiscip. Rev. Data Min. Knowl. Discov. **1**(6), 477–495 (2011)
18. Y.S. Koh, N. Rountree, Finding sporadic rules using apriori-inverse, in *Proceedings of the 9th Pacific-Asia Conference on Advances in Knowledge Discovery and Data Mining, PAKDD'05*, Hanoi, 2005, pp. 97–106
19. N. Lavrač, P.A. Flach, B. Zupan, Rule evaluation measures: a unifying view, in *Proceedings of the 9th International Workshop on Inductive Logic Programming, ILP '99* (Springer, London, 1999), pp. 174–185
20. T. Li, X. Li, Novel alarm correlation analysis system based on association rules mining in telecommunication networks. Inf. Sci. **180**(16), 2960–2978 (2010)
21. W. Li, J. Han, J. Pei, CMAR: accurate and efficient classification based on multiple class-association rules, in *Proceedings of the 1st IEEE International Conference on Data Mining, ICDM 2001*, San Jose, November-December 2001, pp. 369–376
22. B. Liu, W. Hsu, Y. Ma, Integrating classification and association rule mining, in *Proceedings of the Fourth International Conference on Knowledge Discovery and Data Mining, KDD-98*, New York City, August, 1998, pp. 80–86
23. B. Liu, Y. Ma, C.K. Wong, *Classification Using Association Rules: Weaknesses and Enhancements* (Kluwer Academic Publishers, Dordrecht, 2001), pp. 591–601
24. J.M. Luna, J.R. Romero, S. Ventura, G3PARM: a grammar guided genetic programming algorithm for mining association rules, In *Proceedings of the IEEE Congress on Evolutionary Computation, IEEE CEC 2010*, Barcelona, 2010, pp. 2586–2593
25. J.M. Luna, J.R. Romero, S. Ventura, Design and behavior study of a grammar-guided genetic programming algorithm for mining association rules. Knowl. Inf. Syst. **32**(1), 53–76 (2012)
26. J.M. Luna, J.R. Romero, C. Romero, S. Ventura, Reducing gaps in quantitative association rules: a genetic programming free-parameter algorithm. Integr. Comput. Aided Eng. **21**(4), 321–337 (2014)
27. J.M. Luna, J.R. Romero, S. Ventura, On the adaptability of G3PARM to the extraction of rare association rules. Knowl. Inf. Syst. **38**(2), 391–418 (2014)
28. J.M. Luna, C. Romero, J.R. Romero, S. Ventura, An evolutionary algorithm for the discovery of rare class association rules in learning management systems. Appl. Intell. **42**(3), 501–513 (2015)
29. J.M. Luna, A. Cano, M. Pechenizkiy, S. Ventura, Speeding-up association rule mining with inverted index compression. IEEE Trans. Cybern. **46**(12), 3059–3072 (2016)
30. J.M. Luna, F. Padillo, M. Pechenizkiy, S. Ventura, Apriori versions based on mapreduce for mining frequent patterns on big data. IEEE Trans. Cybern. 1–15, 2018. https://doi.org/10.1109/TCYB.2017.2751081
31. J.M. Luna, M. Pechenizkiy, M.J. del Jesus, S. Ventura, Mining context-aware association rules using grammar-based genetic programming. IEEE Trans. Cybern. 1–15, 2018. https://doi.org/10.1109/TCYB.2017.2750919
32. J. Mata, J.L. Alvarez, J.C. Riquelme, Mining numeric association rules with genetic algorithms, in *Proceedings of the 5th International Conference on Artificial Neural Networks and Genetic Algorithms, ICANNGA 2001*, Taipei, 2001, pp. 264–267

33. J. Mata, J.L. Alvarez, J.C. Riquelme, Discovering numeric association rules via evolutionary algorithm, in *Proceedings of the 6th Pacific-Asia Conference on Advances in Knowledge Discovery and Data Mining, PAKDD 2002*, Taipei, 2002, pp. 40–51

34. R. McKay, N. Hoai, P. Whigham, Y. Shan, M. O'Neill, Grammar-based genetic programming: a Survey. Genet. Program Evolvable Mach. **11**, 365–396 (2010)

35. N. Ordoñez, C. Ezquerra, C. Santana, Constraining and summarizing association rules in medical data. Knowl. Inf. Syst. **9**, 1–2 (2006)

36. G. Piatetsky-Shapiro, Discovery, analysis and presentation of strong rules, in *Knowledge Discovery in Databases*, ed. by G. Piatetsky-Shapiro, W. Frawley (AAAI Press, Menlo Park, 1991), pp. 229–248

37. J.R. Quinlan, *C4.5: Programs for Machine Learning* (Morgan Kaufmann, San Mateo, 1993)

38. A. Rahman, C.I. Ezeife, A.K. Aggarwal, Wifi miner: an online apriori-infrequent based wireless intrusion system, in *Proceedings of the 2nd International Workshop in Knowledge Discovery from Sensor Data, Sensor-KDD '08*, Las Vegas, 2008, pp. 76–93

39. D. Sánchez, J.M. Serrano, L. Cerda, M.A. Vila, Association rules applied to credit card fraud detection. Expert Syst. Appl. **36**, 3630–3640 (2008)

40. T. Scheffer, Finding association rules that trade support optimally against confidence, in *Proceedings of the 5th European Conference on Principles of Data Mining and Knowledge Discovery (PKDD 2001)*, Freiburg, 2001, pp. 424–435

41. P. Tan, V. Kumar, Interestingness measures for association patterns: a perspective, in *Proceedings of the Workshop on Postprocessing in Machine Learning and Data Mining, KDD '00*, New York, 2000

42. F.A. Thabtah, A review of associative classification mining. Knowl. Eng. Rev. **22**(1), 37–65 (2007)

43. F. Thabtah, P. Cowling, Y. Peng, MMAC: a new multi-class, multi-label associative classification approach, in *Proceedings of the 4th IEEE International Conference on Data Mining (ICDM'04)*, Brighton, November, 2004, pp. 217–224

44. F. Thabtah, P. Cowling, Y. Peng, MCAR: multi-class classification based on association rule approach, in *Proceedings of the 3rd IEEE International Conference on Computer Systems and Applications*, Cairo, June 2005, pp. 1–7

45. S. Ventura, J.M. Luna, *Pattern Mining with Evolutionary Algorithms* (Springer International Publishing, Cham, 2016)

46. G. Widmer, Learning in the presence of concept drift and hidden contexts. Mach. Learn. **23**(1), 69–101 (1996)

47. X. Yin, J. Han, CPAR: classification based on Predictive Association Rules, in *Proceedings of the 3rd SIAM International Conference on Data Mining, SDM 2003*, San Francisco, May 2003, pp. 331–335

48. C. Zhang, S. Zhang, *Association Rule Mining: Models and Algorithms* (Springer, Berlin, 2002)

49. W. Zhou, H. Wei, M.K. Mainali, K. Shimada, S. Mabu, K. Hirasawa, Class association rules mining with time series and its application to traffic load prediction, in *Proceedings of the 47th Annual Conference of the Society of Instrument and Control Engineers (SICE 2008)*, Tokyo, August 2008, pp. 1187–1192

Chapter 6
Exceptional Models

Abstract A classical task in data analytics is the finding of data subsets that somehow deviate from the norm and denote that something interesting is going on. Such deviations can be measured in terms of the frequency of occurrence (e.g. class association rules) or as an unusual distribution for a specific target variable (e.g. subgroup discovery). Exceptional models, however, identify small subsets of data which distribution is exceptionally different from the distribution in the complete set of data records. In this supervised descriptive pattern mining framework, several target variables are selected, and data subsets are deemed interesting when a model over the targets on the data subset is substantially different from the model on the whole dataset. This chapter formally introduces the task and describes the main differences with regard to similar tasks. Finally, the most important approaches in this field are analysed.

6.1 Introduction

Pattern mining [1, 29] has been defined as a broad subfield of data mining where only a part of the data is described at a time and any type of homogeneity and regularity is denoted for this data subset. In the best-known form of pattern mining, that is frequent itemset mining [1], the interest of a pattern is analyzed in an unsupervised manner (all the items are equally promising beforehand), aiming at finding patterns that frequently occur in data. When patterns are analyzed from a supervised point of view (interesting descriptions of a target variable are required), the most extensively studied task is subgroup discovery [14] (see Chap. 4). Here, the goal is to find subgroups (data subsets) for which the distribution of this target is unusual.

Exceptional model mining [20] (EMM) is defined as a multi-target generalization of subgroup discovery [14], which is the most extensively studied form of supervised descriptive pattern mining [26]. This task searches for interesting data subsets on predefined target variables and describes reasons to understand the cause of the unusual interactions among the targets. Thus, rather than identifying a single attribute as the target t, several target attributes $t_1, \ldots, t_m$ are considered in EMM

© Springer Nature Switzerland AG 2018   129
S. Ventura, J. M. Luna, *Supervised Descriptive Pattern Mining*,
https://doi.org/10.1007/978-3-319-98140-6_6

and the interestingness is evaluated in terms of an unusual joint distribution of $t_1, \ldots, t_m$. The aim is therefore to discover data subsets [19] where the model over the targets on the subset is substantially different from the model on the whole dataset (or the complement of the extracted subset). In other words, given two target attributes t_1 and t_2 and a data subset $S \subseteq \Omega$, the simplest form of EMM calculates the Pearson's standard correlation coefficient between t_1 and t_2 and analyzes the variation in the correlation that makes this subset S to perform an exceptional behavior with respect to Ω (or $\Omega \setminus S$).

EMM works on two main sets: descriptive and target attributes. Descriptive attributes are those responsible for defining the data subset S, whereas the target attributes are used to evaluate S. At this point, and in order to illustrate the type of exceptionality provided by this task, let us take an example widely used in the field about the analysis of the housing price per square meter [9]. Here, the general know-how is that a larger size of the lot coincides with a higher sales price. However, it is of high interest for an investor to know whether it is possible to find specific data subsets where the price of an additional square meter is significantly less than the norm, or even zero. If such subsets can be found, then the speculation can be eased and the benefits increased. For example, it is possible to find out that the price of a house in the higher segments of the market is mostly determined by its location and facilities. The desirable location may provide a natural limit on the lot size, such that this is not a factor in the pricing. Another EMM example is related to the evidence of the Giffen effect in data (the economic law of demand states that the demand of a product will decrease if its price increases) for which there are conditions under which this law does not hold. In [9], authors considered extremely poor households, who mainly consume cheap staple food, as well as relatively rich households in the same neighborhood, who can afford to enrich their meals with a luxury food. In this situation, when the price of the food increases, there will be a point where the relatively rich households can no longer afford the luxury food and they start consuming more of the cheapest food. As a result, an increase in the price of staple food will lead to an increase in the demand for the staple food. This relation, however, does not hold for extremely poor households who can afford less of this food.

A fundamental characteristic of EMM is the finding of reasons to understand why parts of the dataset deviate form the norm. Thus, existing EMM approaches should not only identify data subsets but also describe such subsets. In this regard, a recent research study [23] proposed the extraction of reliable relationships among patterns within a data subset where the analysis on two target variables denotes a significant difference with regard to its complement. Different approaches for EMM were analyzed in a recent research study [9] where it was discovered two main schools of thought in the EMM community. The first set of approaches restricts descriptive attributes in the dataset to be nominal and imposes the anti-monotone constraint [2] so the search space can be explored exhaustively by considering traditional subgroup discovery methods [14]. On the contrary, the second group includes models mainly based on heuristic search approaches, and which enable descriptive attributes to be either discrete and continuous.

6.2 Task Definition

Leman et al. [20] formally described the exceptional model mining (EMM) problem as an extension of subgroup discovery [14], aiming at discovering data subsets where a model fitted to such subsets is substantially different from the same model fitted to the entire database (or the complement) [8]. Let us assume a database Ω formed by a set of records $r \in \Omega$ in the form of $r = \{d_1, \ldots, d_l, t_1, \ldots, t_m\}$: $l, m \in \mathbb{N}^+$. Here, each data record r includes a set of descriptive attributes or descriptors $\{d_1, \ldots, d_l\}$ as well as a set of target variables $\{t_1, \ldots, t_m\}$. EMM aims at discovering a data subset $S \subseteq \Omega$, also denoted as $G_D \subset \Omega$ corresponding to a description given by a set of descriptive attributes $D(d_1, \ldots, d_l)$, satisfying that $G_D = \{\forall r^i \in \Omega : D(d_1^i, \ldots, d_k^i) = 1\}$, and G_D showing an unusual interaction on two specific target variables t_j and t_k. The size of a data subset G_D is denoted as $|G_D|$ and determines the number of records satisfied by D. As for the complement of a subset G_D^C, it is defined as the set of records $G_D^C \subseteq \Omega$ that are not satisfied by D, i.e. $G_D^C = \Omega \setminus G_D$.

The unusual interaction discussed so far should consider two types of comparisions. A possibility is to compare the model for a description D of a subgroup G_D either to the model for its complement $G_D^C \equiv \Omega \setminus G_D$, or to the model for the whole dataset Ω. However, these two options may lead to very different outcomes as it is described below according to a research studied presented by *Duivesteijn et al.* [9]. Let us suppose a dataset Ω comprising 6 records (see Fig. 6.1) that are represented in a two-dimensional target space (t_j and t_k). Here, we are concerned with finding descriptions having a deviating regression line in these two dimensions. The solid grey line r' represents the regression line of the whole dataset (with slope -1), that is, the regression line for the 6 records (3 records depicted as squares and 3 records depicted as circles). Let us suppose now that we find the description D covering those records depicted as circles, producing a data subset G_D with a regression line r'' (dashed grey line) with a slope 1. $G_D^C = \Omega \setminus G_D$ is therefore represented by the 3 squares, having a regression line r''' (dotted line) wth a slope 1. As a result,

Fig. 6.1 Comparison of a data subset G_D with respect to its complement G_D^C or the whole dataset Ω

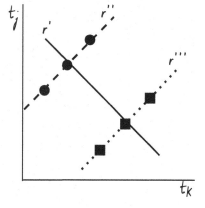

taking into account the difference in the slope of the regression line and considering the sample Ω dataset that was described by *Duivesteijn et al.* in [9], it is obtained that G_D is interesting when compared to the whole dataset Ω. However, G_D is not interesting at all when it is compared to its complement G_D^C.

Taking a dataset with no additional information, it is not always possible to accurately determine whether a data subset G_D should be compared to G_D^C or Ω. When analysing a problem, if the emphasis is placed on deviations from a possibly inhomogeneous norm, then it makes more sense to compare G_D to Ω. On the contrary, if the emphasis is placed on dichotomies, then it makes more sense to compare G_D to its complement G_D^C. On other occasions, a statistically inspired quality measure may require choosing either Ω or G_D^C, to prevent violation of mathematical assumptions. To sum up, it is important for the expert (or end user) to realize that the choice of whether to compare to Ω or G_D^C is essential for the type of outcome. Additionally, not only the outcome is influenced by this choice but also the computational expense. When comparing a dataset Ω to n data subsets $(G_{D_1}, G_{D_2}, \ldots, G_{D_n})$ given a series of descriptions, it is required to learn $n + 1$ models. On the contrary, when comparing n data subsets $(G_{D_1}, G_{D_2}, \ldots, G_{D_n})$ to their complements $(G_{D_1}^C, G_{D_2}^C, \ldots, G_{D_n}^C)$, then $2n$ models are required to be learned.

6.2.1 Original Model Classes

In EMM, the interest of the resulting set G_D given by a set of descriptive attributes $\{d_1, \ldots, d_l\}$ is quantified by a quality measure that assigns a unique numeric value to the description $D(d_1, \ldots, d_l)$ on Ω. The typical quality measure indicates how exceptional the model fitted on the targets in the datas subset is, compared to either the model fitted on the targets in its complement, or the model fitted on the targets in the whole dataset. Originally, three classes of models and suggested quality measures for them were proposed [20].

6.2.1.1 Correlation Models

Given two numeric target variables t_j and t_k, their linear association is measured by the correlation coefficient ρ, which is estimated by the sample correlation coefficient $\hat{r}$ as shown in Eq. (6.1). Here, t_j^i denotes the i-th record in data, and $\overline{t_j}$ denotes the mean of the target variable t_j. ρ_D^G and $\rho_D^{G^C}$ denote the coefficients of correlation for G_D and G_D^C, respectively; whereas $\hat{r}^{G_D}$ and $\hat{r}^{G_D^C}$ denote their sample estimates.

$$\hat{r} = \frac{\Sigma(t_j^i - \overline{t_j})(t_k^i - \overline{t_k})}{\sqrt{\Sigma(t_j^i - \overline{t_j})^2 \Sigma(t_k^i - \overline{t_k})^2}} \tag{6.1}$$

In this type of models, the quality measure is statistically-oriented (based on the test $H_0 : \rho_D^G = \rho^{G_D^C}$ and $H_1 : \rho_D^G \neq \rho^{G_D^C}$) and considers the number of data records and the deviation in the correlation coefficient. If t_j and t_k follow a bivariate normal distribution, the Fisher z transformation can be applied as shown in Eq. (6.2). According to *Neter et al.* [25], the sampling distribution of z' (see Eq. (6.3)) is approximately normal and, as a consequence, z^* follows a standard normal distribution under H_0. Here, z' considers $\hat{r}^{G_D}$, and z'^C uses $\hat{r}^{G_D^C}$. If both $|G_D|$ and $|G_D^C|$ are greater than 25, then the normal approximation is quite accurate and, therefore, the quality measure can be computed as 1 minus the computed p-value for z^*.

$$z' = \frac{1}{2}ln(\frac{1+\hat{r}}{1-\hat{r}}) \tag{6.2}$$

$$z^* = \frac{z' - z'^C}{\sqrt{\dfrac{1}{|G_D|-3} + \dfrac{1}{|G_D^C|-3}}} \tag{6.3}$$

Another consideration for a quality measure in EMM is the absolute difference of the correlation of G_D and its complement $G_D^C \equiv \Omega \setminus G_D$. However, this measure, calculated as $|\hat{r}^{G_D} - \hat{r}^{G_D^C}|$, does not take into account the number of records included in G_D.

6.2.1.2 Regression Models

Considering a regression model $t_j = a + bt_k + e$ fitted to a data subset G_D and its complement G_D^C, and due to the slope b is of primary interest when fitting a regression model, a good quality measure for EMM is the difference in the slopes. It indicates the change in the expected value of t_j when t_k increases. Here, the hypothesis to be tested is $H_0 : b_D^G = b^{G_D^C}$ and $H_1 : b_D^G \neq b^{G_D^C}$. Here, the least squares stimate (see Eq. (6.4)) for the slope b is considered as well as an unbiased estimator s^2 (see Eq. (6.5)) for the variance of $\hat{b}$. Here, e^i is the regression residual for i-th individual, whereas m is the number of records in the subset G_D, or $\overline{m}$ for the number of records in G_D^C. Finally, the test statistic t' can be defined as shown in Eq. (6.6). Similarly to the correlation models, the quality measure is defined as one minus the computed p-value for t' with the degrees of freedom df given by Eq. (6.7). If $m + \overline{m} \geq 40$ the t-statistic is quite accurate, so we should be confident to use it unless we are analysing a very small dataset.

$$\hat{b} = \frac{\Sigma(t_k^i - \overline{t_k})(t_j^i - \overline{t_j})}{\sqrt{\Sigma(t_k^i - \overline{t_k})^2}} \tag{6.4}$$

$$s^2 = \frac{\Sigma(e^i)^2}{(m-2)\,\Sigma(t_k^i - \overline{t_k})^2} \tag{6.5}$$

$$t' = \frac{\hat{b}^{G_D} - \hat{b}^{G_D^C}}{\sqrt{\Sigma(s^{G_D})^2 + (s^{G_D^C})^2}} \tag{6.6}$$

$$df = \frac{((s^{G_D})^2 + (s^{G_D^C})^2)^2}{\dfrac{(s^{G_D})^4}{m-2} + \dfrac{(s^{G_D^C})^4}{\overline{m}-2}} \tag{6.7}$$

6.2.1.3 Classification Models

In classification models, the target variable t_j is defined in a discrete domain whereas descriptive attributes or descriptors $\{d_1, \ldots, d_l\}$ can be of any type (binary, nominal, numeric, etc). In this type of models, any classification method can be used to quantify the exceptionalness. A simple example is the logistic regression model (see Eq. (6.8)) for a binary class label $t_j \in \{0, 1\}$. The coefficient b provides information about the effect of t_k on the probability that t_j occurs. Thus, if b takes a positive value it denotes an increase in t_k leads to an increase of $P(t_j = 1|t_k)$ and vice versa. At this point, given a data subset G_D (D being a set of descriptive attributes $D(d_1, \ldots, d_l)$), the model is fit as $logit(P(t_j = 1|t_k)) = a + b \cdot D + c \cdot t_k + d \cdot (D \cdot t_k)$. Here, authors proposed [20] as quality measure 1 minus the p-value of a standard test in the literature on logistic regression [25] on $d = 0$ against a two-sided alternative (for $D = 1$ and $D = 0$) in the aforementioned model.

$$logit(P(t_j = 1|t_k)) = ln(\frac{P(t_j = 1|t_k)}{P(t_j = 0|t_k)}) = a + b \cdot t_k \tag{6.8}$$

A second possibility is to focus on a simple decision table where the relative frequencies of t_j are computed for each possible combination of values for $\{t_1, \ldots, t_m\} \setminus t_j$. A typical example of this classification model was proposed by *Leman et al.* [20] by describing a hypothetical dataset of 100 people applying for a mortgage and including two attributes describing the age (low, medium, high) and marital status (married, not married). The output determines whether the application was successful or not so the probabilities (in brackets) in per unit basis of having a positive outcome are as follows: *low* $\wedge$ *not married* (0.25); *low* $\wedge$ *married* (0.61); *medium* $\wedge$ *not married* (0.40); *medium* $\wedge$ *married* (0.69); *high* $\wedge$ *not married* (0.73); and *high* $\wedge$ *married* (1.00). According to these results, the classifier predicts a positive outcome in all cases except when *not married* and age is either *low* or *medium*. Here, a possible quality measure to quantified the exceptionalness is based on the *Hillinger distancie* [18], which assigns a value to the distance between the conditional probabilities estimated on G_D and G_D^C. This distancie considers the distribution $P(t_j|t_1, \ldots, t_m)$ for each possible combination $t_1, \ldots, t_m$, and it is calcuated as shown in Eq. (6.9). Here, $\hat{P}^{G_D}$ denotes the probability estimates on G_D, and $\hat{P}^{G_D^C}$ the probability estimates on G_D^C.

$$D(\hat{P}^{G_D}, \hat{P}^{G_D^C}) = \sum_{t_1, \ldots, t_m} \sum_{t_j} (\sqrt{\hat{P}^{G_D}(t_j|t_1, \ldots, t_m)} - \sqrt{\hat{P}^{G_D^C}(t_j|t_1, \ldots, t_m)})^2$$

(6.9)

6.2.2 Rank Correlation Model Class

As described in previous sections, the exceptional interaction among target varaibles can come in many different forms, Pearson's correlation being one of the most straightforward (see Sect. 6.2.1.1). Considering this measure, it is calculated the difference in the linear relation between two targets on a data subset and that same relation on the complement of such a subset. In a recent research study, however, *Duivesteijn et al.* [7] proposed the interaction of two target variables in terms of rank correlation [16]. According to the authors [7], the use of rank correlation models comes with three advantages over the existing correlation model class. First, the rank correlation model does not need the assumption of target normality present in the correlation model. Second, the rank correlation model is less sensitive to outliers. Finally, the gauged form of interaction is richer since it enables to find data subsets where the monotone relation (monotone relations encompass linear relations) between two targets is substantially different from that same relation on the complement of such subsets. All these features are described in depth below.

Focusing on the assumption of normality, it should be stated that 's correlation coefficient implies the assumption that the targets in question are normally distributed. Without the normality assumptions, many statistical tests on Pearson's correlation coefficient become meaningless or hard to interpret. Additionally, real-life examples and datasets cannot guarantee normality and, therefore, the use of Pearson's correleation coefficient is questionable to be applied as a suitable measure in EMM. As a conclusion, it may be asserted that the normality assumption limits the scope of application for the correlation model class (see Sect. 6.2.1.1).

As for the sensitivity to outliers, it has been demonstrated [3] and graphically illustrated (see Fig. 6.2) that Pearson's correlation coefficient is easily affected by outliers. Four different datasets have been considered and they consider almost identical statistical properties (the four sets share the same Pearson's correlation coefficient value). As it is illustrated in Fig. 6.2, there are datasets featuring substantially different relations between the two displayed variables but sharing the same statistical properties.

Finally, focusing on the linear/monotone relation, it should be highlighted that Pearson's correlation focuses on linear relations between two target variables, whereas rank correlation focuses on the richer class of monotone relations between

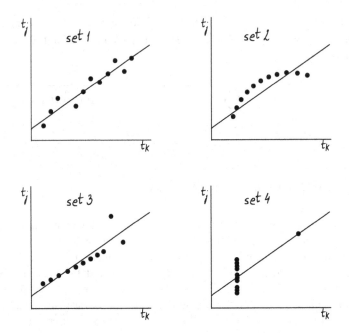

Fig. 6.2 Graphical demonstration of the sensitivity to ouliers proposed by *Anscombe* [3]

two target variables. Hence, EMM with the correlation model class aims at finding data subsets where this linear relation is exceptional, whereas the use of the rank correlation model class proposes the finding of data subsets where the monotone relation between the target variables is exceptional. The correlation model class is more specialized, while the rank correlation model class is more generalized. At this point, it should be remarked that the expert may have domain-specific reasons to prefer one over other.

The rank correlation model class for EMM was formally defined by *Duivesteijn et al.* [7] on the basis of a dataset Ω including a set of records in the form $r = \{d_1, \ldots, d_l, t_x, t_y\} : l \in \mathbb{N}^+$. Here, each data record r includes a set of descriptive attributes or descriptors $\{d_1, \ldots, d_l\}$ as well as two target variables t_x and t_y. The domain for descriptive attributes is unrestricted, whereas for target variables (t_x and t_y) the domain should be numeric or, at least, ordinal (the minimum requirement is to be able to rank the values of t_x and t_y). An example of rank correlation coefficient that uses the difference between rankings of a pair t_x and t_y is Spearman's ρ_s, developed by *C. Spearman* [28] as shown in Eq. (6.10). Here, d_i is the difference between the ranks of t_x and t_y in the i-th record from a total of n records in Ω. If no ties are present, this is equivalent to computing the Pearson's coefficient (see Eq. (6.1)) over the ranks of data (see Eq. (6.11)), considering $R_i^{t_x}$ and $R_i^{t_y}$ as the ranks of t_x and t_y in the i-th records, whereas $\overline{R^{t_x}}$ and $\overline{R^{t_y}}$ denote their respective means. Equation (6.10) should be used in situations where the number of ties is moderate, whereas Eq. (6.11) in case of ties.

$$\rho_s = 1 - \frac{6 \cdot \sum d_i^2}{n \cdot (n^2 - 1)} \qquad (6.10)$$

$$\hat{r} = \frac{\Sigma(R_i^{t_x} - \overline{R^{t_x}})(R_i^{t_y} - \overline{R^{t_y}})}{\sqrt{\Sigma(R_i^{t_x} - \overline{R^{t_x}})^2 \Sigma(R_i^{t_y} - \overline{R^{t_y}})^2}} \qquad (6.11)$$

Another different rank correlation coefficient is given by *Kendall* [16], which defines a statistic based on the agreement (concordances) of ranks to measure the correlation of a sample. Kendall's coeffecent ρ_κ, which makes the correlation measure less sensitive to outliers, is formally defined as shown in Eq. (6.12). In this equation, C is the number of concordants pairs, that is, given a pair of observations (t_x, t_y) and (t_x', t_y') it is concordant if $t_x < t_x' \wedge t_y < t_y'$ or $t_x > t_x' \wedge t_y > t_y'$; D stands for the number of discordants pairs; T_x and T_y are the numbers of pairs tied only on t_x and t_y values, respectively. Thus, a pair is said to be tied if $t_x = t_x'$ or $t_y = t_y'$, and discordant otherwise.

$$\rho_\kappa = \frac{C - D}{\sqrt{(C + D + T_x) \cdot (C + D + T_y)}} \qquad (6.12)$$

Taking into account the two aforementioned rank correlation coefficients, that is, ρ_s and ρ_κ, it is obtained that a good quality measure to for EMM is defined as $\varphi(G_D) = |\rho_s^{G_D} - \rho_s^{G_D^C}|$ when the Spearman's rank correlation coefficient is considered; and $\varphi(G_D) = |\rho_\kappa^{G_D} - \rho_\kappa^{G_D^C}|$ when the Kendall's rank correlation coefficient is bore in mind. Even when these quality measures do not make any assumption on the distribution targets, their major drawback is that the size of the subsets are not considered and it may happen that small subsets that are not interesting at all may obtain extreme correlation values. A solution proposed by *Fieller et al.* [10] is to determine whether the difference in rank correlation on both Kedall's ρ_κ and Spearman's ρ_s are statistically significant by performing a Fisher z-transformation by considering $z_{\rho_s}^{G_D} = arctanh(\rho_s^{G_D})$ or $z_{\rho_\kappa}^{G_D} = arctanh(\rho_\kappa^{G_D})$ so the values are normally distributed. Then, according to *Fieller et al.* [10], they enabled comparisons of Kedall's ρ_κ and Spearman's ρ_s. Fieller-Spearman is denoted in Eq. (6.13), where $var_{\rho_s}(S) = 1.06/(|S| - 3)$ for a subset S. The quality measure for EMM is calculated as 1 minus the computed p-value for $\varphi_{z_{\rho_s}}(G_D)$ (Spearman's rank correlation coefficient).

$$\varphi_{z_{\rho_s}}(G_D) = \frac{z_{\rho_s}^{G_D} - z_{\rho_s}^{G_D^C}}{\sqrt{var_{\rho_s}(G_D) + var_{\rho_s}(G_D^C)}} \qquad (6.13)$$

On the contrary, the rank correlation difference between a data subset and its complement using Fieller-Kendall is denoted as illustrated in Eq. (6.14), where

$var_{\rho_\kappa}(S) = 0.437/(|S| - 4)$ for a subset S. The quality measure for EMM is calculated as 1 minus the computed p-value for $\varphi_{z_{\rho_\kappa}}(G_D)$ (Kendall's rank correlation coefficient).

$$\varphi_{z_{\rho_\kappa}}(G_D) = \frac{z_{\rho_\kappa}^{G_D} - z_{\rho_\kappa}^{G_D^C}}{\sqrt{var_{\rho_\kappa}(G_D) + var_{\rho_\kappa}(G_D^C)}} \qquad (6.14)$$

6.2.3   Related Tasks

Pattern mining aims at describing intrinsic and important properties of data to understand an underlying phenomena without any previous knowledge. Sometimes, though, it is crucial to identify patterns (maybe in the form of comprehensible rules) to describe chunks of data according to a specific target variable, giving rise to the task known as supervised descriptive pattern mining [26] (a specific case of pattern mining). In previous chapters of this book, contrast set mining [6] (see Chap. 2), emerging pattern mining [12] (see Chap. 3), subgroup discovery [14] (see Chap. 4), and class association rule mining [21] (see Chap. 5) were described as important supervised descriptive pattern mining tasks, providing a useful comparison to denote the differences (slight differences sometimes) among all of them.

Let us focus first on the subgroup discovery task, which is the most extensively studied form of supervised descriptive pattern mining [26]. This task was introduced by *Klösgen* [17] and *Wrobel* [30], and it was formally described as follows: given a set of data records and a property of those records that we are interested in, the task of subgroup discovery is to find data subsets that are statistically most interesting for the user, e.g. data subsets that are as large as possible and have the most unusual statistical characteristics with respect to a target variable of interest. Based on this description, *Leman et al.* [20] defined the EMM task as a multi-target generalization of subgroup discovery. An additional key point to be considered is that EMM is more focused on the interactions on two target variables, whereas subgroup discovery is more related to the statistical characteristics of the data subset.

Finally, it is important to remark that, the EMM task is highly related to the *Simpson*'s paradox [24]. This paradox, also known as the *Yule-Simpson* effect, describes that a trend that is present in different data subsets may disappear when these data subsets are combined. A well-known real-world example of the *Simpson*'s paradox was described in [11]. When analysing different applications for admission at the university of California, Berkeley, it was discovered that men were more likely than women to be admitted, and the difference was so large that some remedial actions were required. For instance, 44% of men were admitted, whereas only 35% of women do. Nevertheless, when examining individual departments, it was discovered that no department was significantly biased against women and some of them had small but statistically significant bias in favor of women, i.e. 82% of women were admitted to a specific department whereas only 62% of men do. This unexpected behaviour may also appear in correlations where, similarly to EMM, two variables that appear to have a positive correlation may present a negative correlation when they are analyzed from a different perspective (e. g. the whole dataset). *Berman et al.* [5] described an example that involved an apparent violation of the law of supply and demand, that is, authors described a situation in which price changes seemed to bear no relationship with quantity purchased. This counterintuitive relationship, however, disappeared once the data is divided into subsets of finer time periods.

6.3 Algorithms for Mining Exceptional Models

The EMM task can be divided into three main groups: exceptional model mining in general, exceptional preference mining, and exceptional relationship mining. Each of these groups has its own features that should be bore in mind to design a proper algorithm.

6.3.1 Exceptional Model Mining

The EMM problem has been overcome from different points of view, but there are two main schools of thought in this community. One is related to classical subgroup discovery algorithms [4] where attributes in data are required to be nominal and the search space is explored exhaustively. The other one is related to heuristic search, enabling attributes defined in continuous domains to be considered.

Focusing on exhaustive search approaches, it is important to remark that, due to the problems related to the computational expenses and the imposibility of working on continuous domains (a discretization step is required), no research studies are proposed novela approaches for EMM on exhaustive search methodologies. However, since EMM has been considered as an extension of subgroup discovery by *Leman et al.* [20], it is possible to use adaptations of subgroup discovery approaches. In this regard, exceptional data subsets can be discovered by using the well-known Apriori [2] algorithm or an extension of the Apriori-SD [15] for subgroup discovery. This algorithm (see Algorithm 6.1) is similar to Apriori in the sense that it is based on a level-wise paradigm where all the frequent patterns of length $k+1$ are generated by using all the frequent patterns of length k if and only if they have, at least, $k-1$ items in common. The modified version for EMM includes an additional step (see lines 7 to 14, Algorithm 6.1) where all the previously mined patterns (data subsets represented by such patterns) are analysed on the basis of the target variables t_x and t_y. How exceptional a subset is on the target variables is computed by using either the Spearman's rank correlation or the Kendall's rank correlation. A pattern (a data subset) is considered as of interest if 1 minus the computed p-value is greater than an α threshold.

Similarly to Apriori, EMM can be carried out by considering the well-known FP-Growth algorithm [13]. This algorithm can be modified (see Algorithm 6.2) so

---

Algorithm 6.1 Pseudo-code of the Apriori algorithm for EMM

Require: $\Omega, t_x, t_y, mSupport, \alpha$ ▷ dataset Ω, target variables t_x and t_y, minimum support threshold used to consider the subset, and minimum α value to consider the exceptionalness of a subset

Ensure: S
1: $L_1 \leftarrow \{\forall i \in \Omega : f(i) \geq mSupport\}$ ▷ frequent singletons
2: $C \leftarrow \emptyset$
3: **for** $(k = 1; L_k \neq \emptyset; k{+}{+})$ **do**
4:     $C \leftarrow$ candidates patterns generated from L_k
5:     $L_{k+1} \leftarrow \{\forall c \in C : f(c) \geq mSupport\}$ ▷ frequent patterns of size $k + 1$
6: **end for**
7: **for** $\forall L_k \forall i \in L_k$ **do**   ▷ Analysis of all the discovered frequent patterns or itemsets (data subsets) in order to look for exceptional models on t_x and t_y
8:     Compute G_D based on the pattern or itemset i
9:     Compute G_D^C as $\Omega \setminus G_D$
10:     Compute the value $\varphi_{z_{\rho_s}}(G_D)$ for Spearman's rank correlation (or $\varphi_{z_{\rho_K}}(G_D)$ for Kendall's rank correlation) considering the target variables t_x and t_y
11:     **if** $1 - (p\text{-value of } \varphi_{z_{\rho_s}}(G_D))$ (or $\varphi_{z_{\rho_K}}(G_D)) \geq \alpha$ **then**
12:       $S \leftarrow i$
13:     **end if**
14: **end for**
15: **return** S

Algorithm 6.2 Pseudo-code of the FP-Growth algorithm for EMM

Require: $\Omega, t_x, t_y, mSupport, \alpha$ ▷ dataset Ω, target variables t_x and t_y, minimum support threshold used to consider the subset, and minimum α value to consider the exceptionalness of a subset

Ensure: S
1: $Tree \leftarrow \emptyset$
2: $F\text{-}List \leftarrow \{\forall i \in \Omega : f(i) \geq mSupport\}$ ▷ frequent singletons
3: Sort $F\text{-}List$ in descending order of the frequency
4: Define the root of the tree data representation $Tree$
5: **for all** $r \in \Omega$ **do**   ▷ for each data record in Ω
6:     Make r ordered according to $F\text{-}List$
7:     Include the ordered record r in $Tree$
8: **end for**
9: **for all** $i \in \Omega$ **do**   ▷ for each item in Ω
10:     $P \leftarrow$ Construct any pattern by taking i and traversing the $Tree$ structure
11:     **if** $f(P) \geq mSupport$ **then**   ▷ only frequent patterns are considered
12:       Compute G_D based on the pattern P
13:       Compute G_D^C as $\Omega \setminus G_D$
14:       Compute the value $\varphi_{z_{\rho_s}}(G_D)$ for Spearman's rank correlation (or $\varphi_{z_{\rho_K}}(G_D)$ for Kendall's rank correlation) considering the target variables t_x and t_y
15:       **if** $1 - (p\text{-value of } \varphi_{z_{\rho_s}}(G_D))$ (or $\varphi_{z_{\rho_K}}(G_D)) \geq \alpha$ **then**
16:         $S \leftarrow P$
17:       **end if**
18:     **end if**
19: **end for**
20: **return** S

---

the extracted patterns are then analysed on the basis of two target variables. Here, it should be remarked that a pattern denotes a subset of data and this subset can be easily analysed by using either the Spearman's rank correlation or the Kendall's rank correlation. FP-Growth is much more efficient than Apriori since it works on a compressed data representation in a tree form. In the first step, this algorithm calculates the frequencies of each single item or singleton in Ω, producing a list by including such singletons in descending order of the obtained frequencies. Then, the process of constructing the tree stucture starts from the empty node *null*. The database is scanned for each data record, which is ordered according to the *F-List* and, then every item is added to the tree. If a data record shares the items of any existing branch, then no new branches are needed to be created and the record is inserted in the corresponding path from the root to the common prefix. Finally, once all the data records have been inserted, the tree structure is analysed to obtain any feasible pattern that satisfy a minimum support value (see lines 11 to 18, Algorithm 6.2). For each resulting pattern, the data subset represented by such a pattern is analysed on the basis of two target variables, that is, t_x and t_y. For this analysis, either the Spearman's rank correlation or the Kendall's rank correlation can be considered. A pattern (a data subset) is considered as of interest if 1 minus the computed p-value is greater than an α threshold.

Focusing on heuristic search approaches for the EMM problem, usually the beam search strategy is chosen, which performs a level-wise search. Any beam search algorithm considers a heuristic function f to estimate the cost to reach the goal from a given solution and a beam width w to specify the number of solutions that are stored at each level of a breadth-first search (the w solutions with the best heuristic values at each level of the search). The pseudocode is given in Algorithm 6.3 starting by initializing the *Candidate* set and the S set (the resulting set). After that, the algorithm starts by generating thes best w solutions for a specific heuristic value (φ) at level 1. Thus, in the first level, only descriptions based on singletons are considered. This solution is taken as a candidate solution and all its single descriptors are analyzed, that is, giving a description $d \in D$ comprising $\{d_1, \ldots, d_l\}$, for each d_j a number of descriptions are considered depending on the type (see line 7, Algorithm 6.3): (a) If d_j is binary, then two refined descriptions are considered, that is, one for which D holds and d_j is true, and one for which D holds and d_j is false; (b) If the attribute d_j is nominal with g different values, then $2g$ refined descriptions are considered: for each of the g values, one where D holds and the value is present, and one where D holds and any of the $g - 1$ other values is present; (c) Finally, if the attribute d_j is numeric, we divide the values for d_j that are covered by D into a predefined number b of equal-sized bins. Then, using the $b - 1$ split points $s_1, \ldots, s_{b−1}$ that separate the bins, $2(b - 1)$ refined descriptions are considered: for each split point s_i , one where D holds and d_j is less than or equal to s_i, and one where D holds and d_j is greater than or equal to s_i. For each of this descriptions, it is checked whether they satisfy the constraints or not (see line 10, Algorithm 6.3), which are based on the minimum support value $mSupport$ and the *alpha* value for the quality measure—defined either as Spearman or Kendall (see Sect. 6.2.2) when the rank correlation model is considered.

Algorithm 6.3 Pseudo-code of the beam search algorithm for EMM

Require: $\Omega, t_x, t_y, mSupport, \alpha, w, d$ ▷ dataset Ω, target variables t_x and
   t_y, minimum support threshold used to consider the subset, minimum α value to consider the
   exceptionalness of a subset, beam width w and beam depth d

Ensure: S
  1: $Candidate \leftarrow \emptyset$
  2: $S \leftarrow \emptyset$
  3: **for** $\forall l \in [1, d]$ **do**   ▷ for each level l from 1 till the maximum depth d is reached
  4:     $beam \leftarrow$ Generate the w solutions with the best heuristic values at the level l of the search
  5:     **while** $Candidate \neq \emptyset$ **do**
  6:         $solution \leftarrow$ Take the first solution from $Candidate$
  7:         $D \leftarrow$ Generate a set of descriptions from $solution$
  8:         **for all** $d \in D$ **do**
  9:            Calculate the quality φ of d by considering the target variables t_x and t_y
10:            **if** d satisfies all constraints $(mSupport, \alpha)$ **then**   ▷ d should satisfy the minimum
   support threshold and the α value to consider it as exceptional
11:                $S \leftarrow d$
12:                $beam \leftarrow$ Insert d into $beam$ according its φ
13:            **end if**
14:         **end for**
15:     **end while**
16:     **while** $beam \neq \emptyset$ **do**
17:         $Candidate \leftarrow$ take the first element from $beam$ and remove it from $beam$
18:     **end while**
19: **end for**
20: **return** S

6.3.2 Exceptional Preference Mining

Knobbe et al. [27] introduced the concept of exceptional preferences mining (EPM) as a local pattern mining task that finds data subsets where the preference relations between subsets of the target labels significantly deviate from the norm. It is based on the label ranking concept, that is, a variant of the conventional classification problem where the interest is in assigning a complete preference order of the labels to every example. In EPM, the aim is to look for strong preference behaviour, proposing a measure that denotes that if most data records display a preference $L_1 \succ L_2$, a data subset where most of records display $L_2 \succ L_1$ will be deemed interesting.

In this regard, authors in [27] defined a function w that assigns a value to each pairwise comparison of the labels λ_i and λ_j in such a way that $w(\lambda_i, \lambda_j) = 1$ if $\lambda_i \succ \lambda_j$ (λ_i preferred to λ_j); $w(\lambda_i, \lambda_j) = -1$ if $\lambda_j \succ \lambda_i$; and $w(\lambda_i, \lambda_j) = 0$ if $\lambda_j \sim \lambda_i$. Here, by definition $w(\lambda_i, \lambda_j) = -w(\lambda_j, \lambda_i)$. This function w can also be used to represent a preference matrix that represents sets of rankings from a dataset Ω or data subset $S \subseteq \Omega$. As a matter of example, let us consider the preferences among four labels ($n = 4$) as shown in Table 6.1. A preference matrix M (see Table 6.2) can be produced by considering the mean of w for each pair of labels as $M(i, j) = \sum w(i, j)/n$. In this matrix M, the diagonal will be 0 since $w(i, i) =$

Table 6.1 Sample dataset including the preferences among four labels (attributes of this dataset have been intentially ignored)

Preferences among labels
$\lambda_1 \succ \lambda_2 \succ \lambda_3 \succ \lambda_4$
$\lambda_3 \succ \lambda_1 \succ \lambda_2 \succ \lambda_4$
$\lambda_1 \succ \lambda_2 \succ \lambda_4 \succ \lambda_3$
$\lambda_2 \succ \lambda_3 \succ \lambda_4 \succ \lambda_1$

Table 6.2 Preference matrix obtained from the sample dataset shown in Table 6.1

0.0	0.5	0.0	0.5
−0.5	0.0	0.5	1.0
0.0	−0.5	0.0	0.5
−0.5	−1.0	−0.5	0.0

Table 6.3 Preference matrix obtained from a subset (last two rows) from the sample dataset shown in Table 6.1

0.0	0.0	0.0	0.0
0.0	0.0	1.0	1.0
0.0	−1.0	0.0	0.0
0.0	−1.0	0.0	0.0

0, that is, λ_i compared to itself is 0 (there is no preference). In order to calculate $M(1, 2)$, it is required to compute w resulting as $w(1, 2) = 1$ for the three first records, and $w(1, 2) = -1$ for the fourth record. The average is therefore 0.5, so $M(1, 2) = 0.5$. Analysing the resulting preference matrix M (see Table 6.2), it is easy to detect strong partial order relations in the set since λ_i will be highly preferred if the i-th row has all the values very close to 1. In situations where $M(i, j) = 1$ (or $M(i, j) = -1$) then all the labels in the dataset agree that $\lambda_i \succ \lambda_j$ (or $\lambda_j \succ \lambda_i$).

Once the preference matrix is obtained, it is possible to formally define the quality measures for EPM, denoting how exceptional the preferences are in the whole dataset or the data subsets. A data subset can be considered interesting both by the amount of deviation (distance) and by its size (number of records covered by the data subset), and both measures are normalized to the interval [0, 1]. The size of a data subset S is commonly computed by its square root as $\sqrt{|S|}$ ($|S|$ being the number of records of S), and normalized as $\sqrt{|S|/|\Omega|}$. On the other hand, the distance L between two preference matrices M and M' is computed as $(M - M')/2$ where the data subset that produces M' is a subset of the data subset that produces M. To measure the exceptionality of a subset S, authors in [27] proposed the norm as $Norm(S) = \sqrt{|S|/|\Omega|} \cdot \sqrt{\sum \sum L(i, j)^2}$

As an example, let us consider a subset S of the dataset shown in Table 6.1 by considering only the last two rows (records), that is, $\lambda_1 \succ \lambda_2 \succ \lambda_4 \succ \lambda_3$ and $\lambda_2 \succ \lambda_3 \succ \lambda_4 \succ \lambda_1$. The preference matrix obtained from this subset is illustrated in Table 6.3. Taking both preference matrices (Tables 6.2 and 6.3), the distance between them results as shown in Table 6.4. Finally, the exceptionality is measured as $Norm(S) = \sqrt{0.5} \cdot \sqrt{0.5} = 0.5$.

Table 6.4 Difference matrix
obtained from Tables 6.2 and
6.3

0.0	0.25	0.0	0.25
−0.25	0.0	−0.25	0.0
0.0	0.25	0.0	0.25
−0.25	0.0	−0.25	0.0

6.3.3 Exceptional Relationship Mining

The mining of exceptional relationships [23] among patterns was proposed as a task described halfway between association rule mining [29] and EMM [20]. When the EMM task is performed on different application fields, the discovered information is obtained in the form of patterns that represent a data subset that is somehow exceptional. Nevertheless, and similarly to the pattern mining task [1], the set of patterns obtained by exceptional models does not describe reliable relations between items within the model class. In this regard, the mining of exceptional relationships was proposed [23] as a really interesting task that plays an important role in the knowledge discovery process. This new task, a subtask or extension of EMM, enables the discovery of accurate relations among patterns where the data subset fitted to a specific set of target features is significantly different to its complement. These relationships provide a much more rich and descriptive information than the one obtained by traditional EMM.

The task of mining exceptional relationship is considered as a special procedure with some common properties from both EMM and association rule mining. In order to describe the importance of this task, let us consider exceptional pattern $P = \{d_1, d_2\}$ that represents a data subset given the descriptions d_1 and d_2. This data subset that was obtained thanks to the mining of exceptional models, identifies only a exceptional behaviour in data so it does not provide any descriptive behaviour among its items, that is, between descriptors d_1 and d_2. In such a way, association rule mining can provide a better description of the extracted exceptional model, being able to identify strong associations within the model in the form IF d_1 THEN d_2. This rule describes that it is highly probable that description d_2 appears once d_1 has already appeared. Hence, the main difference between exceptional models and exceptional relationships relies on the fact that the former describes exceptional data subsets comprising a conjunction of descriptors, whereas exceptional relationships form rules that describe strong implications between descriptors within each exceptional data subset. Additionally, it should be noted that the computational cost of mining exceptional relationships is higher than the one of EMM since it is required the mining of both exceptional subgroups and association rules within each exceptional data subset. Finally, the exceptional relationships mining task is formally described as follows. Let us consider a dataset Ω comprising a set of data records $r \in \Omega$, and a data subset G_D defined on the basis of a set of descriptions $D = \{d_1, \ldots, d_l\}$. The mining of exceptional relationships is defined as the task of finding all the set of association rules R with corresponding minimal relationship description, such that each $R_i \in R$ describes a reliable behaviour within the exceptional data subset G_D.

Similarly to any pattern mining algorithm, the extraction of exceptional relationships can be approached in two main forms: exhaustive search methodologies [1] and evolutionary approaches [29]. To date, however, this recent task has been only approached by means of evolutionary computation [23] due to its ability to work on continuous domains and on huge search spaces. Additionally, since the extraction of exceptional relationships in the intersection between exceptional models and association rules, any approach in this field inherits the drawbacks considered by any association rule mining algorithm (it has been already demonstrated [29] that evolutionary computation provides really interesting advantages such as resctriction of the search space, adding subjective knowledge to the mining process, and dealing with different kind of patterns, among others).

The first approach for mining exceptional relationships was based on a grammar-guided genetic programming methodology as it was described by *Luna et al.* [23]. This evolutionary algorithm, known as MERG3P (Mining Exceptional Relationships with Grammar-Guided Genetic Programming), defines a grammar to mine association rules in the form of class association rules [22], that is, solutions comprise a set of items (descriptors) in the antecedent of the rule and only a single item (descriptor) in the consequent. The proposed grammar G (see Fig. 6.3) enables both discrete and continuous descriptors to be obtained, and only a simple modification in the grammar provokes the extraction of any type of rules (a better description about the use of grammars in class association rules is provided in Chap. 5). MERG3P encodes each solution as a sentence of the language generated by the grammar G (see Fig. 6.3), considering a maximum number of derivations to avoid extremely large solutions. Thus, considering the grammar defined in this approach, the following language is obtained $L(G) = \{ (AND\ Condition)^n\ Condition \rightarrow Condition : n \geq 0 \}$. To obtain individuals, a set of derivation steps is carried out by applying the production rules declared in the set P and starting from the start symbol *Rule*.

$G = (\Sigma_N, \Sigma_T, P, S)$ with:

    S = Rule
    Σ_N = {Rule, Conditions, Consequent, Condition, Condition_Nominal,
         Condition_Numerical }
    Σ_T = {'AND', 'Attribute', 'Consequent_value', '=', 'IN','Min_value','Max_value',
         'Consequent_attribute' }
    P = {Rule = Conditions, Consequent ;
         Conditions = 'AND', Conditions, Condition | Condition ;
         Condition = Condition_Nominal | Condition_Numerical ;
         Condition_Nominal = 'Attribute', '=', 'value' ;
         Condition_Numerical = 'Attribute', 'IN', 'Min_value', 'Max_value' ;
         Consequent = Condition_Nominal | Condition_Numerical ; }

Fig. 6.3 Context-free grammar to represent exceptional association rules expressed in extended BNF notation

At this point, it is interesting to remark that exceptional relationships are obtained from an exceptional data subset on two targert variables t_x and t_y. Hence, each solution provided by MERG3P includes some additional features not included in the rule. MERG3P encodes solutions comprising either a rule represented by means of the context free grammar (see Fig. 6.3), and a string of bits where the i-th bit represents the i-th item within the dataset. In this string of bits, a value 1 in the i-th bit means that the i-th item is used as a target feature, where a value 0 means that it is not considered and satisfying that only two bits can take the value 1 at time.

MERG3P starts by generating a set of solutions through the aforementioned context-free grammar (see line 5, Algorithm 6.4) and assigning two random target variables t_x and t_y to the obtained pattern (association rule). Such target variables are defined in a continuous domains so at least two numeric variables are required to be included in Ω. Once this set is obtained, then the evolutionary process starts and a series of processes are carried out over a specific number of generations (see lines 7 to 14, Algorithm 6.4). In a first process, a tournament selector is considered to select a subset of solutions from the set *population* to act as parents (see line 8, Algorithm 6.4), and two genetic operators are then applied to this set of parents. The first genetic operator is responsible for creating new solutions by randomly generating new conditions from the set *parents*. In this regard, this specific genetic operator randomly chooses a condition from the set of conditions of a given individual or solution, creating a completely new condition that replaces the old-one. Then, the second genetic operator is applied by grouping individuals based on their target variables t_x and t_y. Those individuals that were defined for the same

---

Algorithm 6.4 Pseudo-code of the MERG3P algorithm.

Require: Ω, *population_size*, *maxGenerations*, *mSupport*, *mConf*   ▷ dataset, maximum number of solutions to be found, maximum number of generations, and mininimum support and confidence values

Ensure: *population*

1: *population* $\leftarrow \emptyset$
2: *parents* $\leftarrow \emptyset$
3: *offspring* $\leftarrow \emptyset$
4: *number_generations* $\leftarrow 0$
5: *population* $\leftarrow$ generate *population_size* with random target variables t_x and t_y from Ω
6: Evaluate each solution within *population* by considering Ω and the thresholds *mSupport* and *mConf*
7: **while** *number_generations* $<$ *maxGenerations* **do**
8:     *parents* $\leftarrow$ select parents from *population*
9:     *offspring* $\leftarrow$ apply genetic operators to *parents*
10:    Evaluate each solution within *offspring* by considering Ω and the thresholds *mSupport* and *mConf*
11:    *aux_population* $\leftarrow$ *offspring* $\cup$ *popluation*
12:    *population* $\leftarrow$ select the best *population_size* solutions from *aux_population*
13:    *number_generations* $+ +$
14: **end while**
15: **return** *population*

---

pair of targets are clustered in the same group. Then, the process takes the worse individual within each group and modified the target variables so the individual is moved to a different group.

The evaluation process is another major issue in MERG3P (see line 6 and line 10, Algorithm 6.4), since it is required to quantify the exceptionalness of each rule by assigning a value through a fitness function. This evalution process is based on correlation coefficient defined in Eq. (6.1) (see Sect. 6.2.1.1). The proposed fitness function is based on three equations, that is, the correlation coefficient as well as support and confidence quality measures (these two measures are quality measures defined for association rule mining as it was described in Chap. 5). In situations where a certain individual (in the form of an association rule) does not satisfy the predefined quality value for support and confidence, then a 0 fitness function value is assigned to this solution. On the contrary, the fitness value is determined by the absolute difference between the correlation coefficient of the data subset G_P given by the pattern P (the individual that form the association rule) for some target features and the correlation coefficient of its complement G_P^C, that is $|\hat{r}(G_D) - \hat{r}(G_D^C)|$. Therefore, starting from an individual (in the form of an association rule R), the fitness function F used in this algorithm is the one defined in Eq. (6.15).

$$
F(R) = \begin{cases} |\hat{r}(G_D) - \hat{r}(G_D^C)| & \begin{array}{l} \text{if } mSupport \leq Support(R) \quad \wedge \\ Confidence(R) \geq mConf \end{array} \\ \\ 0 & \text{otherwise} \end{cases} \tag{6.15}
$$

The final step of each generation of the evolutionary process is the updating of the set *population*. In this regard, the best *population_size* individuals according to the fitness function value are taken to form the new population. In order to avoid repeated individuals, if two individuals are the same (considering both the rule and the target variables), then only one is considered. It may happen that two individuals represent the same rule but they are defined on different target variables. In such a situation, they are considered dissimilar individuals.

Finally, it is interesting to analyse the search space for this problem, which is higher than the one of EMM in two target variables as well as the one of mining just association rules (without considering the exceptionality). In many situations, the number of both target variables and rules associated to that targets could be extremely high and the problem can be considered as computationally intractable. In this regard, the use of an evolutionary algorithm like MERG3P is highly recommended since exhaustive search approaches may suffer from both computational and memory complexity. It may justify why the problem of mining exceptional relationships has not been addressed from an exhaustive search methodology.

References

1. C.C. Aggarwal, J. Han, *Frequent Pattern Mining* (Springer, Berlin, 2014)
2. R. Agrawal, T. Imielinski, A.N. Swami, Mining association rules between sets of items in large databases, in *Proceedings of the 1993 ACM SIGMOD International Conference on Management of Data, SIGMOD Conference '93, Washington, DC, USA* (1993), pp. 207–216
3. F.J. Anscombe, Graphs in statistical analysis. Am. Stat. **27**(1), 17–21 (1973)
4. M. Atzmueller, Subgroup discovery - advanced review. WIREs: Data Min. Knowl. Disc. **5**, 35–49 (2015)
5. S. Berman, L. DalleMule, M. Greene, J. Lucker, Simpson's paradox: a cautionary tale in advanced analytics, The Statistics Dictionary, 25 September 2012
6. G. Dong, J. Bailey (eds.), *Contrast Data Mining: Concepts, Algorithms, and Applications* (CRC Press, West Palm Beach, 2013)
7. L. Downar, W. Duivesteijn, Exceptionally monotone models - the rank correlation model class for exceptional model mining. Knowl. Inf. Syst. **51**(2), 369–394 (2017)
8. W. Duivesteijn, A.J. Knobbe, A. Feelders, M. van Leeuwen, Subgroup discovery meets bayesian networks – an exceptional model mining approach, in *Proceedings of the 2010 IEEE International Conference on Data Mining, ICDM 2010, Sydney, Australia* (IEEE Computer Society, Washington, 2010), pp. 158–167
9. W. Duivesteijn, A. Feelders, A.J. Knobbe, Exceptional model mining - supervised descriptive local pattern mining with complex target concepts. Data Min. Knowl. Disc. **30**(1), 47–98 (2016)
10. E.C. Fieller, H.O. Hartley, E.S. Pearson, Tests for rank correlation coefficients. i. Biometrika **44**(4), 470–481 (1957)
11. D. Freedman, R. Pisani, R. Purves, *Statistics*, 4th edn. (W. W. Norton, New York, 2007)
12. A.M. García-Vico, C.J. Carmona, D. Martín, M. García-Borroto, M.J. del Jesus, An overview of emerging pattern mining in supervised descriptive rule discovery: taxonomy, empirical study, trends and prospects. Wiley Interdiscip. Rev. Data Min. Knowl. Disc. **8**(1) (2018)
13. J. Han, J. Pei, Y. Yin, R. Mao, Mining frequent patterns without candidate generation: a frequent-pattern tree approach. Data Min. Knowl. Disc. **8**, 53–87 (2004)
14. F. Herrera, C.J. Carmona, P. González, M.J. del Jesus, An overview on subgroup discovery: foundations and applications. Knowl. Inf. Syst. **29**(3), 495–525 (2011)
15. B. Kavšek, N. Lavrač, APRIORI-SD: adapting association rule learning to subgroup discovery. Appl. Artif. Intell. **20**(7), 543–583 (2006)
16. M. Kendall, J.D. Gibbons, *Rank Correlation Methods. A Charles Griffin Title* (E. Arnold, London, 1990)
17. W. Klösgen, Explora: a multipattern and multistrategy discovery assistant, in *Advances in Knowledge Discovery and Data Mining*, ed. by U.M. Fayyad, G. Piatetsky-Shapiro, P. Smyth, R. Uthurusamy (American Association for Artificial Intelligence, Menlo Park, 1996), pp. 249–271
18. L. Le Cam, G. Lo Yang, *Asymptotics in Statistics: Some Basic Concepts*. Springer Series in Statistics (Springer, Berlin, 2000)
19. M. Leeuwen, A. Knobbe, Diverse subgroup set discovery. Data Min. Knowl. Disc. **25**(2), 208–242 (2012)
20. D. Leman, A. Feelders, A.J. Knobbe, Exceptional model mining, in *Proceedings of the European Conference in Machine Learning and Knowledge Discovery in Databases, Antwerp, Belgium*. ECML/PKDD 2008, vol. 5212 (Springer, Berlin, 2008), pp. 1–16
21. B. Liu, W. Hsu, Y. Ma, Integrating classification and association rule mining, in *Proceedings of the Fourth International Conference on Knowledge Discovery and Data Mining, KDD-98, New York City, New York, USA* (1998), pp. 80–86
22. J.M. Luna, J.R. Romero, S. Ventura, Design and behavior study of a grammar-guided genetic programming algorithm for mining association rules. Knowl. Inf. Syst. **32**(1), 53–76 (2012)

23. J.M. Luna, M. Pechenizkiy, S. Ventura, Mining exceptional relationships with grammar-guided genetic programming. Knowl. Inf. Syst. **47**(3), 571–594 s(2016)
24. Y.Z. Ma, Simpson's paradox in GDP and per capita GDP growths. Empir. Econ. **49**(4), 1301–1315 (2015)
25. J. Neter, M.H. Kutner, C.J. Nachtsheim, W. Wasserman, *Applied Linear Statistical Models* (Irwin, Chicago, 1996)
26. P.K. Novak, N. Lavrač, G.I. Webb, Supervised descriptive rule discovery: a unifying survey of contrast set, emerging pattern and subgroup mining. J. Mach. Learn. Res. **10**, 377–403 (2009)
27. C. Rebelo de Sá, W. Duivesteijn, C. Soares, A.J. Knobbe, Exceptional preferences mining, in *Proceedings of the 19th International Conference on Discovery Science DS 2016, Bari, Italy* (2016), pp. 3–18
28. C. Spearman, The proof and measurement of association between two things. Am. J. Psychol. **15**, 88–103 (1904)
29. S. Ventura, J.M. Luna, *Pattern Mining with Evolutionary Algorithms* (Springer, Berlin, 2016)
30. S. Wrobel, An algorithm for multi-relational discovery of subgroups, in *Proceedings of the 1st European Symposium on Principles of Data Mining and Knowledge Discovery, PKDD '97, London, UK* (Springer, Berlin, 1997), pp. 78–87

Chapter 7
Other Forms of Supervised Descriptive Pattern Mining

Abstract Patterns are sometimes used to describe important properties of different data subsets previously identified or labeled, transforming therefore the pattern mining concept into a more specific one, *supervised descriptive pattern mining*. In general, supervised descriptive discovery is said to gather three main tasks: contrast set mining, emerging pattern mining and subgroup discovery. In previous chapters, two additional tasks (class association rules and exceptional models) have been added to this group of three main tasks due to their huge interest in recent years. Nevertheless, there are many additional tasks that can be seen as specific cases of supervised descriptive patterns. In this regard, the aim of this chapter is to provide formal definitions for these tasks, describing their main differences with regard to those already defined in previous chapters.

7.1 Introduction

Supervised descriptive pattern mining includes tasks aiming at providing any kind of descriptive knowledge from labeled data [14]. Techniques proposed for these tasks usually seek descriptors or concepts that characterize and/or discriminate a target variable (or multiple targets). Traditional tasks in this regard are contrast set mining, emerging pattern mining and subgroup discovery. In any of these tasks, the aim is to describe important data subsets according to a specific target or targets of interest. These three tasks have been grouped under the concept supervised descriptive rule discovery [28]. Nowadays, however, different authors are proposing new forms of supervised descriptive patterns [11] and new tasks (somehow related to the three aforementioned traditional tasks) are being considered as methods in the supervised descriptive pattern mining field: exceptional models and class association rules. In fact, the mining of class association rules provides a knowledge that can be used in any of the existing supervised descriptive pattern mining tasks as it is described in Chap. 5.

In this chapter, the aim is to extend the number of tasks defined under the term *supervised descriptive pattern mining*. In this regard, the first example is relative to the representation of compacted descriptors, that is, a reduced set

© Springer Nature Switzerland AG 2018
S. Ventura, J. M. Luna, *Supervised Descriptive Pattern Mining*,
https://doi.org/10.1007/978-3-319-98140-6_7

that formally reveals the same information as the whole set, as it is the case of closed patterns [1]. Similarly to unlabeled data, closed patterns can be used to compress information from labeled data, which is really useful for classification and discrimination purposes [15]. Another important form of looking for supervised descriptive patterns is bump hunting [12], which looks for subregions of the space of variables within which the value of the target variable is considerably larger (or smaller) than its average value over the whole search space. Bump hunting is therefore a specific case of subgroup discovery [23] where the interest is not focused on metrics of complexity, generality, precision, and interest. Instead, this task aims at mining data subsets that highly characterize a target variable (defined in a discrete domain). Additionally, bump hunting can be seen as a specific case of exceptional models or relationships [25] with the difference that the former analyse the behaviour on a single target variable whereas the latter considers a pairs of target variables.

Another example of task is impact rule mining [39], which is defined as a specific type of class association rules where the consequent is a target variable defined in a continuous domain. Unlike class association rules, impact rules express the consequent together with different measures including coverage, mean, variance, maximum value, minimum value, among others. Thus, an impact rule is represented as a class association rule in the form $X \rightarrow Target\{coverage = V_1, mean = V_2, variance = V_3, maximum\_value = V_4, ...\}$ where V_1 is the percentage of records in which the set X is true; V_2 is the mean of values of the $Target$ variable for which X is true; V_3 is the variance of the $Target$ values for which X is true; etc.

When considering data mining techniques for dealing with historical decision records, there is a lack of care about minorities or protected-by-law groups. On this matter, *Turini et al.* [31] proposed the notion of discrimination discovery as a criterion to identify and analyse the potential risk of discrimination. In this problem, given a dataset of historical decision records, a set of potentially discriminated groups, and a criterion of unlawful discrination, the aim is to find subsets of the decision records, called context, and potentially discriminated groups within each context for which the criterion of unlawful discrimination hold. In discrimination discovery, a key concept is considered to expresses the relative variation in the reliability of the rule due to the addition of an extra itemset in the antecedent of the base rule. This tasks is highly related to the extraction of context-aware patterns [26], and relations among patterns.

The extraction of class association rules [40] from black box models is another example of supervised descriptive pattern mining. This task aims at revealing the hidden knowledge of such models, for example those obtained from neural networks, as well as to describe the process they follow to produce a final decision (output) [4]. Finally, another important and different form of supervised descriptive pattern mining addressed in this chapter is known as change mining. It aims at modelling, monitoring, and interpreting changes in those patterns that describe an evolving domain over many temporally ordered data sets [7]. The final objective of this task is to produce knowledge that enables any anticipation of the future so it is particularly helpful in domains in which it is crucial to detect emerging trends as early as possible and which require to make decisions in anticipation rather than in reaction.

All the aforementioned tasks are briefly described in this chapter so the aim is to provide the reader with additional tasks that can be considered when seeking descriptors that characterize a target (or multiple targets) variable. It should be highlight that the aim is not to provide any existing task but a set of tasks, so many additional approaches can be found in the literature related to the supervised descriptive pattern mining field.

7.2 Additional Tasks

According to *Novak et al.* [28], some additional research studies are closely related supervised descriptive pattern mining. Therefore, in addition to the tasks described in previous chapters it is possible to list the following tasks.

7.2.1 Closed Sets for Labeled Data

Any of the existing tasks for mining supervised descriptive patterns aim at seeking a space of concept descriptions for labeled data (a target variable is considered in this regard). When analysing this space, some descriptions may turn out to be more relevant than others for characterizing and discriminating the target variable. In descriptive data mining [37], relevant descriptions in the form of patterns were provided by closure systems [8] aimed at compacting the whole set of descriptions into a reduced set that formally conveys the same information. For example, condensed representations of relevant patterns were proposed as solutions to overcome existing problems in the pattern mining field (those relative to computational and storage issues) [1], easing as well the final knowledge discovery process.

Garriga et al. [15] first introduced the use of such condensed representations on the space of concept descriptions for labeled data. Here, authors considered a set of descriptors $D = \{d_1, ..., d_n\}$ in a dataset Ω as features representing attribute-value pairs. The target variable or class label cannot be included in the set of features so given an example, it corresponds to a subset of features in D with an associated class label that is not included in D. As a matter of clarification, let us consider exactly the same example used by *Garriga et al.* [15], which based on the well-known contact lens dataset (see Table 7.1). In this sample dataset, patients are described by four different attributes (age, prescription, astigmatism and tear production rate) and a target variable (type of lens). When focusing on a specific target value, all the data records that are labeled with such a value are denoted as positive examples, whereas examples of all the other values are treated as negative examples.

Table 7.1 Contact lens dataset proposed in [9]

Age	Prescription	Astigmatism	Tear production	Lens
Young	Myope	No	Normal	Soft
Young	Hypermetrope	No	Normal	Soft
Pre-presbyopic	Myope	No	Normal	Soft
Pre-presbyopic	Hypermetrope	No	Normal	Soft
Presbyopic	Hypermetrope	No	Normal	Soft
Young	Myope	No	Reduced	None
Young	Myope	Yes	Reduced	None
Young	Hypermetrope	No	Reduced	None
Young	Hypermetrope	Yes	Reduced	None
Pre-presbyopic	Myope	No	Reduced	None
Pre-presbyopic	Myope	Yes	Reduced	None
Pre-presbyopic	Hypermetrope	No	Reduced	None
Pre-presbyopic	Hypermetrope	Yes	Reduced	None
Pre-presbyopic	Hypermetrope	Yes	Normal	None
Presbyopic	Myope	No	Reduced	None
Presbyopic	Myope	No	Normal	None
Presbyopic	Myope	Yes	Reduced	None
Presbyopic	Hypermetrope	No	Reduced	None
Presbyopic	Hypermetrope	Yes	Reduced	None
Presbyopic	Hypermetrope	Yes	Normal	None
Young	Myope	Yes	Normal	Hard
Young	Hypermetrope	Yes	Normal	Hard
Pre-presbyopic	Myope	Yes	Normal	Hard
Presbyopic	Myope	Yes	Normal	Hard

Based on the aforementioned issues, closed frequent patterns [1] (see Sect. 1.2.3, Chap. 1) were considered to obtain condensed representations on the space of concept descriptions for labeled data. Closed frequent patterns can be formally defined as those frequent patterns belonging to the set $\mathscr{P}$, which includes all the feasible patterns, that have no frequent superset with their same frequency of occurrence, i.e. $P_i, P_j \in \mathscr{P}, \nexists P_j \supset P_i : f(P_j) = f(P_i)$. From the practical point of view, closed patterns are the largest sets among those other patterns occurring in the same data records. Considering class label *soft* (see Table 7.1), all data records with this label are considered as positive examples, and all data records labeled *none* or *hard* are considered as negative. In this subset of positive examples, the itemset $\{young\}$ is not a closed pattern since there is a superpattern $\{young, no, normal\}$ with the same support in this subset, that is, $f(\{young\}) = f(\{young, no, normal\}) = 2$. Thus, considering all the positive examples of a dataset, it is possible to extract any feasible closed frequent pattern and represent them in a Hasse diagram (see Fig. 7.1 for positive examples on the class label *soft*). In this graphical organization, each node corresponds to a closed pattern (frequency

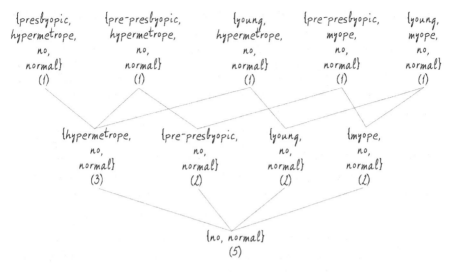

Fig. 7.1 Hasse diagram of closed patterns for class label *soft* from Table 7.1

of occurrence is denoted in brackets), and there is an edge between two nodes if and only if they are comparable and there is no other intermediate closed pattern in the lattice. Here, ascending/descending paths represent the superset/subset relation. Finally, it is important to remark that, when talking about frequent closed patterns it is required to consider a minimum support (frequency) constraint given by the user.

Garriga et al. [15] defined that a set of descriptors $D' \subseteq D$ belonging to the set of descriptors $D = \{d_1, ..., d_n\}$ in data is more relevant than the set $D'' \subseteq D$ if and only if all the positive data records (those including the target value) covered by D'' are also covered by D', and all the negative data records (those not including the target value) covered by D' are also covered by D''. It can be immediately demonstrated that if D' is more relevant than D'' in the positives examples, then D'' will be more relevant than D' in the negatives. Additionally, given two different closed sets $D' \subseteq D$ and $D'' \subseteq D$ on those data records satisfying the target value such as $D' \not\subseteq D''$ and $D'' \not\subseteq D'$, then these sets do not share any path in the lattice (Hasse diagram) and, therefore, they cannot be compared in terms of relevancy since they cover different data records. As a matter of clarification, let us consider the lattice illustrated in Fig. 7.1 and two closed sets $\{young, no, normal\}$ and $\{myope, no, normal\}$. Here, they cover different data records for the target value *soft* so they cannot be compared in terms of relevance.

Following wth the analysis, given two closed sets $D' \subseteq D$ and $D'' \subseteq D$ on those data records satisfying the target such as $D' \subset D''$, it is obtained that the set of positive data records (those including the target value) covered by D' is a subset (not equal) of those covered by D''. Similarly to previous definitions, all the negative data records (those not including the target value) covered by D'' are also covered by D'. It points out that two different closed sets may cover exactly the same set of negative data records, and it is preferable the one covering a higher number of positive data records (the anti-monotonicity property of support determines that the smaller pattern will be the most relevant [1]). To illustrate this assertion, let us consider the same lattice of closed itemsets (see Fig. 7.1). The closed set $\{myope, no, normal\}$ covers fewer positives data records (considering the feature $soft$ as class label) than the proper predecessor $\{no, normal\}$ and both sets cover exactly the one negative data record (those records not including the feature $soft$), so $\{no, normal\}$ is more relevant than $\{myope, no, normal\}$.

Considering all the aforementioned definitions and taking the well-known contact lens dataset (see Table 7.1) with a target variable (type of lens), it is interesting to obtain all the relevant closed sets for the value $soft$ of such target variable. According to Fig. 7.1, the following ten closed patterns are considered on this target value: $\{presbyopic, hypermetrope, no, normal\}$, $\{pre\text{-}presbyopic,$ $hypermetrope, no, normal\}$, $\{young, hypermetrope, no, normal\}$, $\{pre\text{-}$ $presbyopic, myope, no, normal\}$, $\{young, myope, no, normal\}$, $\{hypermetrope,$ $no, normal\}$, $\{young, no, normal\}$, $\{pre\text{-}presbyopic, no, normal\}$, $\{myope, no,$ $normal\}$, and $\{no, normal\}$. From this set, let us consider just those closed patterns that cover more than a single positive data record. From this reduced set, the closed pattern $\{no, normal\}$ covers a higher number of positive records but a higher number of negative records so those closed patterns covering a lower number of negative records are denoted as relevant (they do not share any path). Finally, the closed pattern $\{no, normal\}$ covers the same number of negative records as the closed pattern $\{myope, no, normal\}$, but the latter covers a lower number of positive records and they are in the same path. Thus, $\{no, normal\}$ is more relevant than $\{myope, no, normal\}$. As a result, the resulting set of closed patterns is reduced to four relevant closed patterns: $\{no, normal\}$, $\{hypermetrope, no,$ $normal\}$, $\{pre\text{-}presbyopic, no, normal\}$, and $\{young, no, normal\}$.

Finally, when considering the class label $hard$, a total of 7 closed patterns are obtained and this set is reduced to only 3 relevant closed patterns. On the other hand, the class label $none$, a total of 61 closed patterns are discovered, which is reduced to 19 relevant closed patterns.

7.2.2 Bump Hunting

An important form of looking for supervised descriptive patterns is bump hunting [12], which looks for subregions of the space of features within which the value of the target variable is considerably larger (or smaller) than its average value over the whole search space. There are many problems where finding such regions is of considerable practical interest. As an example, let us suppose the problem of loan acceptance faced by a bank, where people with a low risk of defaulting are preferable to those with a high risk. In order to obtain the risk default, different characteristics of the applicant are considered such as income, age, occupation, and so on. The bank may be therefore interested in analysing the collected data in the past in order to group applicants with a low probability of defaulting, analysing their characteristics and clarifying whether or not to accept future applicants.

In a formal way, let us considered a collection of input variables or features $F = \{f_1, f_2, ..., f_n\}$, and a variable of interest (target variable) denoted by t. Let us also denote S_j as the set of possible values of f_j. The input space is therefore denoted as $S = S_1 \times S_2 \times ... \times S_n$. The aim of bump hunting is to find subregions R of such input space S, that is, $R \subset S$ for which $\overline{t_R} \gg \overline{t}$. Here $\overline{t}$ is the global mean of the target variable t, whereas $\overline{t_R}$ states for the mean of t in the subregion R. The aim is therefore to extract subregions where the mean of the target variable is much greater than the mean of the whole dataset. It is important to remark that, in the formal definition [12], the input variables as well as the target variable may be either categorical or numeric. In case that the target variable is defined in a categorical domain with two different values, then it is most convenient to code two different classes as 0 and 1, so that the average value of t may be interpreted as the probability that $t = 1$. On the contrary, if t is defined as a categorical variable including more than two values, then the problem cannot be handled directly and it is required to reformulate it as a collection of binary problems (one class with value 1 and the rest with label 0).

In bump hunting, authors named the desired subregions as boxes due to their rectangular shape. Let us consider $s_j \subseteq S_j$ as a feasible value s_j of f_j (as it was previously denoted, S_j is the set of all possible values of f_j). A box b_i is formally defined as $b_i = s_1 \times s_2 \times ... \times s_n$. In situations where $s_j = S_j$, that is, f_j takes any value in its domain, then f_j can be removed from the box. An example of a box b_1 is illustrated in Fig. 7.2, including two features defined in continuous domains, that is, $b_1 = f_1 \in [v_1^{f_1}, v_2^{f_1}] \wedge f_2 \in [v_1^{f_2}, v_2^{f_2}]$. Another example is illustrated in Fig. 7.3, showing a sample b_2 that includes two features defined in discrete domains, that is, $b_2 = f_3 \in \{v_2^{f_3}, v_3^{f_3}\} \wedge f_4 \in \{v_2^{f_4}\}$. A final example (see Fig. 7.4) illustrates a box

Fig. 7.2 Example of a numeric box

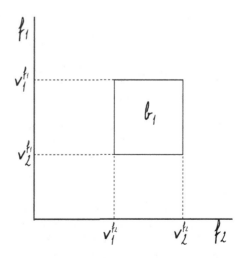

Fig. 7.3 Example of a
categorical box

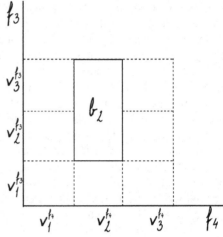

Fig. 7.4 Example of a box
including categorical and
numeric features

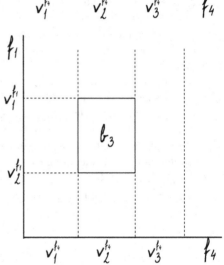

b_3 that includes two features, one defined in a discrete domain and the other one
defined in a continuous domian, that is, $b_3 = f_1 \in [v_1^{f_1}, v_2^{f_1}] \wedge f_4 \in \{v_2^{f_4}\}$.

At this point, it is important to remark that, even when the original definition
considered box construction, it is equivalent to rule induction (rules in the form
$f_j \wedge ... \wedge f_k \rightarrow t$) where the target variable t is used just to compute the
difference between the mean $\overline{t_R}$ in the subset and the mean $\overline{t}$ in the whole dataset.
It demonstrates that this task is highly related to other supervised descriptive
pattern mining tasks such as subgroup discovery (see Chap. 4) and class association
rule mining (see Chap. 5). Additionally, since it looks for subsets where a target
variable differently behave with regard to the whole dataset, it is also related to the
exceptional model mining task (see Chap. 6).

One of the simplest forms of mining box (in the form of rules) in bump hunting
is through a covering strategy. This strategy can be performed sequentially by using

---

Algorithm 7.1 Pseudo-code of the strategy to form boxes in bump hunting

---

Require: $\Omega, t, mSupport, mMean$ ▷ dataset Ω, target variable t, minimum support and
     mininum mean value for the target t
Ensure: *box*   ▷ best box found
1: $L_1 \leftarrow \{\forall i \in \Omega : f(i \wedge t) \geq mSupport\}$ ▷ frequent singletons by considering the target
    variable
2: $C \leftarrow \emptyset$
3: **for** $(k = 1; L_k \neq \emptyset; k + +)$ **do**
4:     $C \leftarrow$ candidates patterns generated from L_k
5:     $L_{k+1} \leftarrow \{\forall c \in C : f(c \wedge t) \geq mSupport\}$ ▷ frequent patterns of size $k + 1$ by
    considering the target variable
6: **end for**
7: $P \leftarrow \cup_k L_k$ ▷ set of candidate boxes that satisfy the mininimum support value
8: Compute the mean $\overline{t_{p_j}}$ for all $p_j \in P$
9: $box \leftarrow$ Take the best candidate p_j from P according to the difference between $\overline{t_{p_j}}$ and $\overline{t_\Omega}$
10: **return** *box*

Algorithm 7.2 Pseudo-code of the top-down refinement for the PRIM algorithm

---

Require: $\Omega, B, t, mSupport$ ▷ dataset Ω, box B to be peeling, target t and minimum support
Ensure: B
1: **repeat**
2:     $C \leftarrow$ generate the set of candidates for removal
3:     $b \leftarrow$ take the subbox whose mean for the target t is maximum with regard to the mean for
    the general B
4:     $B \leftarrow B \setminus b$
5: **until** $support(B) \leq mSupport$
6: **return** B

---

an exhaustive search strategy to subsets of data (see Algorithm 7.1). The first box is constructed on the whole dataset, whereas the second box is obtained by removing the data records that fall into the first box and applying the same strategy. Box construction continues until there is no box in the remaining data that includes a minimum number of data records (minimum support value) and sufficiently high target mean. It is important to highlight that this algorithm can be performed only on discrete features (or numeric features if they are previously discretized). Besides, this approach turns computationally intractable for large number of features and any kind of heuristic search is required.

One of the most important algorithms in bump hunting is called PRIM (Patient Rule Induction Method) [12], which consists of a main phase based on a top-down refinement. This refinement is related to the top-down peeling (see Algorithm 7.2), taking a box B that covers all the remaining data in each specific moment. A small subset (subbox b) within the general box B is chosen (to be removed) by considering the one that yields the largest target value within $B \setminus b$. The aim is to refine the box B by adding constraints (features). It can be considered as a hill-climbing algorithm since at each step in the search, the aim is to peel of the single subbox that gives the most improvement of the target mean. The procedure is repeated till the support of the current box is lower than a predefined minimum support value $mSupport$.

Table 7.2 Sample bank credit data

Married	Owner	Gender	Loan acceptance
No	No	Male	0
No	Yes	Female	0
Yes	Yes	Male	0
No	No	Male	0
Yes	Yes	Female	0
Yes	Yes	Female	1
Yes	Yes	Male	1
Yes	No	Male	1
No	Yes	Female	1
Yes	Yes	Female	1

Target variable is the loan acceptance (0 or 1 to represent, no or yes respectively)

As a matter of clarification, let us consider the remaining data shown in Table 7.2 in which we are interested in searching for groups with a high mean of the target variable (loan acceptance). In other words, our interest is in those features that represent groups with high probability of having a loan accepted. Let us also consider a minimum support threshold value of 0.4 (no boxes below this value can be accepted) and the target mean of the whole dataset is 0.5 as illustrated in Table 7.2. On this dataset, a number of peeling actions can be performed: it is possible to remove those records comprising the feature $married = no$ resulting $\bar{t} = 0.666$; another option is to remove those records comprising the feature $married = yes$ resulting $\bar{t} = 0.250$; if those records with $owner = yes$ are removed, it results in $\bar{t} = 0.333$; removing thosd records with $owner = no$ results in $\bar{t} = 0.571$; not considering those records with $gender = male$ obtains $\bar{t} = 0.600$; finally, it is also possible to remove those records comprising the feature $gender = female$ resulting $\bar{t} = 0.400$. From all of these posibilities, the one that leads to highest mean in the remaining box is the one that removes the feature $married = no$. The resulting box (see Table 7.3) has a support of 0.6 which is above the minimum support threshold $mSupport$ (predefined to the value 0.4). The peeling process is repeated by considering a set of candidate actions to be performed: it is possible to remove those records comprising the feature $owner = no$ resulting $\bar{t} = 0.833$; another possibility is to remove those records comprising the feature $owner = yes$ resulting $\bar{t} = 1.000$; a third possibility is to remove those records comprising the feature $gender = male$ resulting $\bar{t} = 0.666$; finally, not considering the data records including the feature $gender = female$ results in $\bar{t} = 0.666$. From all these possibilities, the one with the highest mean is $owner = yes$ but it gives rise to a dataset comprising a single record (support value $0.1 \leq mSupport$) so it cannot be taken into account. Thus, the best option is to remove those data records comprising the feature $owner = no$ (support value $0.5 > mSupport$). Considering the remaing data records, no more iterations can be

Table 7.3 Resulting box
after removing records with
the feature *married* = *no*
from the original data (see
Table 7.2)

Owner	Gender	Loan acceptance
Yes	Male	0
Yes	Female	0
Yes	Female	1
Yes	Male	1
No	Male	1
Yes	Female	1

Algorithm 7.3 Pseudo-code of the Beam Search algorithm

Require: $\Omega, B, t, mSupport$ ▷ dataset Ω, box B to be peeling, target t and minimum support
Ensure: *BeamSet*
 1: *BeamSet* ← *B*
 2: **repeat**
 3:     *all* − *subboxes* ← ∅
 4:     **for** ∀B_i ∈ *BeamSet* **do**
 5:         C_i ← generate the set of candidates subboxes satisfying *mSupport*
 6:         *all* − *subboxes* ← *all* − *subboxes* ∪ C_i
 7:     **end for**
 8:     *BeamSet* ← take the best *w* subboxes from *all* − *subboxes*
 9: **until** no improvement is possible or a number of repetitions is done
10: **return** *BeamSet*

---

peformed since any additional actions will not satisfy the *mSupport* value. The final
box is defined as IF *married* = *yes* ∧ *owner* = *yes* THEN *Loan acceptance* = 1
(*support* = 0.5).

An alternative algorithm for bump hunting is based on a Beam Search (see
Algorithm 7.3), also called Data Surveyor [18]. Similarly to the PRIM [12]
algorithm, it begins with a box *B* that covers all the data and a number of subboxes
$b_1, ..., b_w$ within the initial box *B* are selected. These *w* subboxes are chosen
as follows. First, the mean value of the target variable in the subbox should be
significantly higher than the mean in the current box. Second, it is required that
the resulting subbox satisfies at least a fraction of data records, that is, its support
should be higher than a predefined minimum value *mSupport*. Finally, from the
candidate subboxes, those *w* subboxes with the largest mean are chosen. Notice that
at each level in the search we only consider a total of *w* subboxes of the boxes at
the previous level, rather than the *w* best subboxes of each box at the previous level.
Comparing this approach to the PRIM algorithm, they are similar except for the
selection of the *w* best subboxes (PRIM only takes the best one). Thus, if *w* = 1
then both algorithms are the same.

Taking again the sample dataset shown in Table 7.2, the Beam Search algorithm
with *w* = 3 will take the three best peeling actions: *married* = *no* producing
$\bar{t}$ = 0.666; *gender* = *male* obtaining $\bar{t}$ = 0.600; and *owner* = *no* producing
$\bar{t}$ = 0.571. From them, a new set of candidate subboxes are produced from which

the best $w = 3$ ones are considered. Considering *married* $= no$ the following candidates to remove records are obtained:

- *owner* $= no$ with $\bar{t} = 0.600$.
- *owner* $= yes$ with $\bar{t} = 1.000$.
- *gender* $= male$ with $\bar{t} = 0.666$.
- *gender* $= female$ with $\bar{t} = 0.666$.

Considering *gender* $= male$ the following candidates are obtained:

- *owner* $= no$ with $\bar{t} = 0.600$.
- *owner* $= yes$ with $\bar{t} = 0.000$.
- *married* $= no$ with $\bar{t} = 0.666$.
- *married* $= yes$ with $\bar{t} = 0.500$.

Finally, considering *owner* $= no$ the following candidates are obtained:

- *married* $= no$ with $\bar{t} = 0.600$.
- *married* $= yes$ with $\bar{t} = 0.500$.
- *gender* $= male$ with $\bar{t} = 0.600$.
- *gender* $= female$ with $\bar{t} = 0.500$.

From these set of 12 candidate subboxes, the best $w = 3$ options are the following: *owner* $= no$ (for the previous *married* $= no$) and *owner* $= no$ (for the previous *gender* $= male$). Notice that it is not possible to take more than two solutions since the remain candidates do no satisfy the minimum support threshold ($mSupport = 0.4$). The final boxes are defined as IF *married* $= yes \wedge owner = yes$ THEN *Loan acceptance* $= 1$ (*support* $= 0.5$), and IF *gender* $= female \wedge owner = yes$ THEN *Loan acceptance* $= 1$ (*support* $= 0.5$).

7.2.3 Impact Rules

Assoiation rules [40] have been widely used to denote any interesting association between items or elements in data. Most of the existing approaches for association rule discovery [3] are based on discrete data and there is limited research into the best way of discovering rules from quantitative data [24]. In fact, first research studies [40] discretized the quantitative variables and mapped them into qualitative (discrete or categorical) ones. Nevertheless, qualitative data have a lower level of measurement scale than quantitative data. Simply applying descretization may lead to information loss [1].

Aumann et al. [5] proposed a variant of association rules whose consequent is quantitative, and it is described by its distribution instead of being discretized into categorical attributes. Authors called these type of rules as quantitative association rules, but in order to differentiate them from the definition provided by *Srikant et al.* [34], *Webb et al.* [39] considered them as impact rules. Considering the

---

Algorithm 7.4 Pseudo-code of the OPUS_IR algorithm

Require: $currentRule, C, Constraints$ ▷ current rule, set C of candidate conditions, and set of constraints specified by the users
Ensure: $RuleList$
  1: $RuleList \leftarrow \emptyset$
  2: $SoFar \leftarrow \emptyset$
  3: **for** $\forall c_i \in C$ **do**
  4:     $New \leftarrow currentRule \cup c_i$
  5:     **if** New satisfies all the constraints in $Constraints$ **then**
  6:         $RuleList \leftarrow RuleList \cup New$
  7:         OPUS($New, SoFar, Constraints$)
  8:         $SoFar \leftarrow SoFar \cup c_i$
  9:     **end if**
10: **end for**
11: **return** $RuleList$

---

formal definition provided in [20], an impact rule consists fo a conjunction of conditions (antecedent of the rule) associated with a numeric value called the target. The consequent of the rule comprises one or more statistics describing the impact on the target caused by the set of data records that satisfy the antecedent. In other words, an impact rule takes the form $antecedent \rightarrow t$, where the target t is described by the following measures: $sup, mean, var, max, min, sum$ and $impact$. Here, sup denotes the percentage of data records satisfied by $antecedent$; $mean$ (and var) states for the mean (and variance) of the target t on those data records satisfied by $antecedent$; max and min represent, respectively, the maximum and minimum values of the target t on those data records satisfied by $antecedent$; sum is the sumatory of the values of the target t on those data records satisfied by $antecedent$; finally, $impact$ is an interesting measure suggested in [39] and defined as $impact(antecedent \rightarrow t) = (mean(antecedent \rightarrow t) - \bar{t}) \times support(antecedent)$. Here, $mean(antecedent \rightarrow t)$ denotes the mean of the target values covered by $antecedent$, whereas $\bar{t}$ is the overall mean of the target values in the whole dataset. A sample impact rule could be $married = yes \wedge profession = programmer \rightarrow income(coverage = 0.45; mean = 57,500; variance = 3,750; max = 85,000; min = 42,000; sum = 2,180,000; impact = 4,112.32)$.

An important algorithm for mining impact rules is known as OPUS_IR (Optimized Pruning for Unordered Search on Impact Rules) [19] which is based on an efficient branch and bound search algorithms [38]. OPUS_IR (see Algorithm 7.4) succesfully overcomes the efficiency problems by performing a search space pruning and a rule discovery in a single phase. The algorithm is initially called with the arguments $currentRule = \emptyset$ and C including the set af all available conditions in a dataset Ω. It should be noted that the set of constraints specified by the users are related to the statistics included in the consequent of the rule, that is, support, mean, variance, maximum and minimum values, impact, etc. The original algorithm in which OPUS_IR is based was developed to provide efficient search for combination of elements in order to optimize a specific metric. Considering a dataset with four

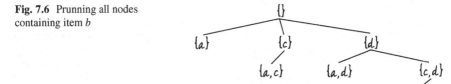

Fig. 7.5 Fixed-structure search space for 4 elements

Fig. 7.6 Prunning all nodes
containing item b

elements (a, b, c and d) and considering each combination of elements just once,
then the search space comprises $2^n - 1 = 2^4 - 1 = 15$ elements as shown in Fig. 7.5.
It is obvious that this search space (for large problems) can only be explored if it is
pruned. In this regard, it is required to prune any node containing the element to be
pruned (for example, the item b) as it is illustrated in Fig. 7.6.

7.2.4 Discrimination Discovery

Data mining techniques for dealing with historical decision records may cause a
discrimination or unequal treatment of minorities or protected-by-law groups. On
this matter, *Turini et al* [31] proposed the notion of discrimination discovery as a
criterion to identify and analyse the potential risk of discrimination. In this problem,
given a dataset of historical decision records, a set of potentially discriminated
groups, and a criterion of unlawful discrination, the aim is to find subsets of the
decision records, called context, and potentially discriminated groups within each
context for which the criterion of unlawful discrimination hold.

   In the simplest form, potentially discriminated groups can be obtained by
specifying a set of selected attribute values (female gender, ethnic minority, etc)
as potentially discriminatory. Sometimes, discrimination can be the result of several
features acting at time that are not discriminatory in isolation. To understand it,

Turini et al. [31] proposed a really interesting example related to a black cat crossing your path, which is well-established as a sign of bad luck. However, this issue is independent of being a cat, being black or crossing a path (these features in isolation are not discriminatory). Additionally, the conjunction of two potentially discriminatory itemsets is a potentially discriminatory itemset as well.

In a formal way, discrimination discovery is performed through the extraction of class association rules [40] of the form $X_1 \wedge X_2 \rightarrow y$; where X_1 and X_2 are itemsets defined in data, whereas y is the target variable or class attribute. Such rules, obtained from a dataset of historical decision records, are denoted as potentially discriminatory rules (hereinafter referred to as PDR) when they include pontentially discriminatory itemsets X_1 in their premises. In order to measure the discrimination power of a PDR, the notion of α-protection is introduced to determine the relative gain in confidence of the rule due to the presence of the discriminatory itemsets. As a matter of clarification, let us consider the same example provided in [31], that is, the rules $city = NYC \rightarrow credit = bad$ and $race = black \wedge city = NYC \rightarrow credit = bad$. The first rule, with a confidence value of 0.25, describes that people who live in NYC (New York City) are usually assigned with a bad credit 25% of the time. The second rule, on the contrary, describes that black people wotho live in NYC are usually assigned with a bad credit 75% of the time (confidence value for this rule was 0.75). As it is demonstrated, the additional item (discriminatory item $race = black$) in the antecedent makes the confidence of the rule up to three times. α-protection is intended to detect rules where such an increase is lower than a fixed threshold α. The discrimination discovery problem can also be performed as an indirect discrimination problem, which is much more challenging. Now, rules are extracted from a dataset where potentially discriminatory itemsets, such as $race = black$, are not present. Continuing with the same example, let us consider the rule $neighborhood = 10451 \wedge city = NYC \rightarrow credit = bad$ (confidence value 0.95). Taking this rule in isolation it is not possible to consider whether it is discriminatory or not. However, it is possible to analyze whether people from neighborhood 10451 are in majority black, that is, it is possible to search for the rule $neighborhood = 10451 \wedge city = NYC \rightarrow race = black$. If the reliability (confidence value) of such a rule is high enough, it is possible to infer the rule $race = black \wedge neighborhood = 10451 \wedge city = NYC \rightarrow credit = bad$ from the previous ones. This inferred rule has a confidence value of 0.94, which is 3.7 times the value obtained by the rule $city = NYC \rightarrow credit = bad$.

In discrimination discovery, a key concept (known as extended lift) is considered to expresses the relative variation of confidence due to the addition of the extra itemset X_2 in the antecedent of the base rule $X_1 \rightarrow y$. In a formal way, the extended lift is defined as let $X_1 \wedge X_2 \rightarrow y$ be a class association rule such that the confidence value of the rule $X_1 \rightarrow y$ is greater than 0. The extended lift of the rule with respect to X_1 is defined as $confidence(X_1 \wedge X_2 \rightarrow y)/confidence(X_1 \rightarrow y)$. X_1 is named the context, and $X_1 \rightarrow y$ the base-rule. In general terms, the extended lift ranges over $[0, \infty)$ except in such situations wherea minimum support value $minSup$ is considered, producing a extended lift that ranges over $[0, 1/minSup]$.

Similarly, if a minimum confidence value $minConf$ is considered for the base-rules, then the extended lift ranges over $[0, 1/minConf]$.

Pedreschi et al. [30] proposed a series of quality measures to determine the differences in the outcomes between two groups in terms of the proportion of people in each group with a specific outcome. Here, it was defined the risk difference (RD), also known as absolute risk reduction, as $RD = confidence(X_1 \wedge X_2 \rightarrow y) - confidence(X_1 \wedge \overline{X_2} \rightarrow y)$. Another metric was defined as the ratio or relative risk ($RR = confidence(X_1 \wedge X_2 \rightarrow y)/confidence(X_1 \wedge \overline{X_2} \rightarrow y)$). Relative chance (RC), also known as selection rate, was denoted as $RD = (1 - confidence(X_1 \wedge X_2 \rightarrow y))/(1 - confidence(X_1 \wedge \overline{X_2} \rightarrow y))$. Odds ration (OR) defined as $OR = (confidence(X_1 \wedge X_2 \rightarrow y) \times (1 - confidence(X_1 \wedge \overline{X_2} \rightarrow y)))/(confidence(X_1 \wedge \overline{X_2} \rightarrow y) \times (1 - confidence(X_1 \wedge X_2 \rightarrow y))$. Finally, a series of measures when the group is compared to $X_1 \wedge X_2 \rightarrow y$ is compared to $X_1 \rightarrow y$: extended difference ($ED = confidence(X_1 \wedge X_2 \rightarrow y) - confidence(X_1 \rightarrow y)$); extended lift ($EL = confidence(X_1 \wedge X_2 \rightarrow y)/confidence(X_1 \rightarrow y)$); and extended chance ($EC = (1 - confidence(X_1 \wedge X_2 \rightarrow y))/(1 - confidence(X_1 \rightarrow y))$).

7.2.5 Describing Black Box Models

The extraction of class association rules [40] from black box models, for example those obtained from neural networks [29], is an aproximation that tend to reveal the hidden knowledge of such models as well as to describe the process they follow to produce a final decision (output) [4]. Focusing on neural networks and in order to make them more human-comprehensible, different research studies have been focussed on extracting rules and describing their behaviour (or the behaviour of their components) in the form of rules. Existing methods to extract descriptive rules can be categorized into three main groups depending on the relationship between the extracted rules and the architecture of the neural network [4]: decompositional approaches, pedagogical approaches, and eclectic approaches.

Decompositional approaches extract rules from neural networks by working on neuron-level, that is, these methods usually analyse each neuron and form rules to copy its behaviour. All the rules extracted for each unit level are then grouped to form a general set to describe the neural network as a whole. Since its definition, many different researchers have proposed approaches based on a decompositional strategy. One of the first methods was named KT^3 [13], which analyses every layer within the network as well as every neuron within each layer through IF-THEN rules by heuristically searching for combinations of input attributes. An improvement of KT^3 was proposed by *Tsukimoto* [36]. This approach also uses a layer-by-layer decompositional method to extract IF-THEN rules for every single neuron but this process is performed a polynomial way (the computational complexity of KT^3 was exponential). Finally, another interesting models based on decompositional is proposed by *Sato et al.* [32]. This algorithm, known as CRED (Continuous/discrete

Rule Extractor via Decision tree induction), transforms each output unit of a neural network into a decision tree where internal nodes of the tree are used to test the hidden layer's units, whereas leave nodes denote the final class (the output).

Pedagogical approaches do not consider the internal structure of the neural networks. Instead, these approaches extract rules by directly mapping inputs and outputs [33]. In other words, these approaches only have access to the output of the neural network for an arbitrary input (no insights into the neural network's structure is considered), looking for coherences between variations of the input and outputs. One of the first methodologies in this sense considered a sensitivity analysis to extract rules in such a way that input intervals should be adjusted to produce a stable output, that is, the predicted class should be the same. Different pedagogical approaches [21] have been based on creating an extensive artificial training set as a basis for the rule learning. Rules can be generated through this sampled set of artificial data in order to imitate the behaviour of the neural network.

Finally, eclectic approaches are based on features of both pedagogical and decompositional approaches [33]. According to *Andrews et al.* [4], eclectic approaches utilise knowledge about the internal architecture and/or weight vectors in the neural network to complement a symbolic learning algorithm. Some of the most important approaches in this category are based on the idea of MofN rules instead of IF-THEN rules. MofN rules [27] are true if at least M of a set of N specified features are present in an example. An example is the MofN algorithm [35] which extracts MofN rules from a neural network by clustering the inputs of a neuron in equivalence classes (similar weights). Then, rules are formed using the clusters.

7.2.6   Change Mining

Change mining is the final task (related to supervised descriptive pattern mining) to be considered. According to its formal definition [7], change mining aims at modelling, monitoring, and interpreting changes in the patterns that describe an evolving domain over many temporally ordered data sets. The final objective of this task is to produce knowledge that enables any anticipation of the future so it is particularly helpful in domains in which it is crucial to detect emerging trends as early as possible and which require to make decisions in anticipation rather than in reaction. In this regard, a typical real application example for this task is the analysis of customer and business processes, that is, a business which operates in dynamic markets with customers who have highly changing demands and attitudes. In such a business, it is a keypoint to understand the customers' expectations since they are generally driven by innovations and competing products.

Let a dataset Ω including a set of items $I = \{i_1, i_2, ..., i_n\}$ that represent binary concepts, that is, each single item may or may not appear in data, taking the value 1 or 0, respectively. Let $T = \{t_1, t_2, ..., t_m\}$ be the collection of transactions or data records in Ω satisfying $\forall t \in T : t \subseteq I \in \Omega$. A change for a pattern $P \subseteq I$, which

Fig. 7.7 Analysis of the
history of support and
confidence along the time

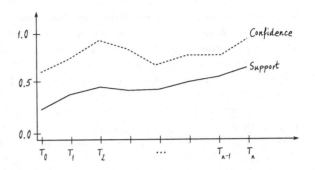

consists of conjunctions of items from I, is quantified as somekind of variation when P is considered in another dataset (a different time period). In this regard, change mining can be seen by many authors [10] as a specific case of contrast set mining (previously defined in Chap. 2) or even emerging patterns (previously defined in Chap. 3). Contrast mining methods [6, 17] usually take two datasets and compare them based on the patterns (models) contained in them, denoting and describing what has changed, in terms of differences. However, these methods are unable to answer how something changes over time [7], giving rise to the field known as change mining (study of time-associated data).

There are many approaches for mining changes that are based on frequent patterns, and almost all these approaches aim at detecting interesting changes by analyzing the support, confidence, or any other measure of interest [16, 37] along the time axis. Figure 7.7 illustrates this analysis where a time-stamped dataset is partitioned into intervalos along the time axis where different quality measures (support and confidence in this example) vary their values along the time. The use of frequent pattern mining [1] can be applied to each of these subsets (defined by the time) in order to obtain histories of the quality measures for each pattern and detect regularities in the histories which are called change patterns.

Change mining based on frequent patterns is a time series analysis problem, but classical approaches in this field cannot directly be applied to pattern histories [7]. The first change mining approach based on frequent patterns was proposed by *Agrawal et al.* [2] in which the user has to specify several shapes (different levels of interest) and the histories are then matched against them. In a different work, *Liu et al.* [22] considered the interest of an association rule not by its quality measures (support, confidence or whatever [16]) but according to its behavior over time (if it exhibits a trend, stability, etc).

References

1. C.C. Aggarwal, J. Han, *Frequent Pattern Mining* (Springer, Berlin, 2014)
2. R. Agrawal, G. Psaila, Active data mining, in *Proceedings of the 1st ACM International Conference on Knowledge Discovery and Data Mining, SIGKDD 95, CA, USA* (1995), pp. 3–8

3. R. Agrawal, T. Imielinski, A.N. Swami, Mining association rules between sets of items in large databases, in *Proceedings of the 1993 ACM SIGMOD International Conference on Management of Data, SIGMOD Conference '93, Washington, DC, USA* (1993), pp. 207–216

4. R. Andrews, J. Diederich, A.L. Tickle, Survey and critique of techniques for extracting rules from trained artificial neural networks. Knowl.-Based Syst. **8**(6), 373–389 (1995)

5. Y. Aumann, Y. Lindell, A statistical theory for quantitative association rules, in *Proceedings of the 5th ACM SIGKDD International Conference on Knowledge Discovery and Data Mining, KDD '99, San Diego, California, USA* (1999), pp. 261–270

6. S.D. Bay, M.J. Pazzani, Detecting group differences: mining contrast sets. Data Min. Knowl. Disc. **5**(3), 213–246 (2001)

7. M. Boettcher, Contrast and change mining. WIREs Data Min. Knowl. Disc. **1**(3), 215–230 (2011)

8. C. Carpineto, G. Romano, *Concept Data Analysis: Theory and Applications* (Wiley, London, 2004)

9. J. Cendrowska, PRISM: an algorithm for inducing modular rules. Int. J. Man Mach. Stud. 349–370 (1987)

10. G. Dong, J. Bailey, (eds.), *Contrast Data Mining: Concepts, Algorithms, and Applications* (CRC Press, Boca Raton, 2013)

11. W. Duivesteijn, A. Feelders, A.J. Knobbe, Exceptional model mining - supervised descriptive local pattern mining with complex target concepts. Data Min. Knowl. Disc. **30**(1), 47–98 (2016)

12. J.H. Friedman, N.I. Fisher, Bump hunting in high-dimensional data. Stat. Comput. **9**(2), 123–143 (1999)

13. L. Fu, Rule generation from neural networks. IEEE Trans. Syst. Man Cybern. **24**(8), 1114–1124 (1994)

14. A.M. García-Vico, C.J. Carmona, D. Martín, M. García-Borroto, M.J. del Jesus, An overview of emerging pattern mining in supervised descriptive rule discovery: taxonomy, empirical study, trends and prospects. WIREs Data Min. Knowl. Disc. (In Press)

15. G.C. Garriga, P. Kralj, N. Lavrač, Closed sets for labeled data. J. Mach. Learn. Res. **9**, 559–580 (2008)

16. L. Geng, H.J. Hamilton, Interestingness measures for data mining: a survey. ACM Comput. Surv. **38**, 9 (2006)

17. R.J. Hilderman, T. Peckham, A statistically sound alternative approach to mining contrast sets, in *Proceedings of the 4th Australasian Data Mining Conference (AusDM), Sydney, Australia* (2005), pp. 157–172

18. M. Holsheimer, Data surveyor: searching the nuggets in parallel. in *Advances in Knowledge Discovery and Data Mining* (1996), pp. 447–467

19. S. Huang, G.I. Webb, Discarding insignificant rules during impact rule discovery in large, dense databases, in *Proceedings of the 2005 SIAM International Conference on Data Mining, SIAM 2005, Newport Beach, CA, USA* (2005), pp. 541–545

20. S. Huang, G.I. Webb, Pruning derivative partial rules during impact rule discovery, in *Proceedings of the 9th Pacific-Asia Conference on Knowledge Discovery and Data Mining, Hanoi, Vietnam.* PAKDD 2005, vol. 3518 (2005), pp. 71–80

21. U. Johansson, T. Lofstrom, R. Konig, C. Sonstrod, L. Niklasson, Rule extraction from opaque models– a slightly different perspective, in *Proceedings of the 5th International Conference on Machine Learning and Applications, ICMLA 2006, Orlando, Florida, USA* (2006), pp. 22–27

22. B. Liu, Y. Ma, R. Lee, Analyzing the interestingness of association rules from the temporal dimension, in *Proceedings of the 1st IEEE International Conference on Data Mining, ICDM 2001, San Jose, CA, USA* (2001), pp. 377–384

23. J.M. Luna, J.R. Romero, C. Romero, S. Ventura, On the use of genetic programming for mining comprehensible rules in subgroup discovery. IEEE Trans. Cybern. **44**(12), 2329–2341 (2014)

24. J.M. Luna, J.R. Romero, C. Romero, S. Ventura, Reducing gaps in quantitative association rules: a genetic programming free-parameter algorithm. Integr. Comput. Aided Eng. **21**(4), 321–337 (2014)

25. J.M. Luna, M. Pechenizkiy, S. Ventura, Mining exceptional relationships with grammar-guided genetic programming. Knowl. Inf. Syst. **47**(3), 571–594 (2016)

26. J.M. Luna, M. Pechenizkiy, M.J. del Jesus, S. Ventura, Mining context-aware association rules using grammar-based genetic programming. IEEE Trans. Cybern. 1–15 (2017). https://doi.org/10.1109/TCYB.2017.2750919

27. P.M. Murphy, M.J. Pazzani, ID2-of-3: constructive induction of M-of-N concepts for discriminators in decision trees, in *Proceedings of the 8th International Workshop on Machine Learning, Evanston, IL, USA* (1991), pp. 183–187

28. P.K. Novak, N. Lavrač, G.I. Webb, Supervised descriptive rule discovery: a unifying survey of contrast set, emerging pattern and subgroup mining. J. Mach. Learn. Res. **10**, 377–403 (2009)

29. M. Paliwal, U.H. Kumar, Neural networks and statistical techniques: a review of applications. Exp. Syst. Appl. **36**(1), 2–17 (2009)

30. D. Pedreschi, S. Ruggieri, F. Turini, A study of top-k measures for discrimination discovery, in *Proceedings of the 27th Symposium on Applied Computing, SAC 2012, Riva del Garda, Italy* (ACM, New York, 2012)

31. S. Ruggieri, D. Pedreschi, F. Turini, Data mining for discrimination discovery. ACM Trans. Knowl. Discov. Data **4**(2), 1–40 (2010)

32. M. Sato, H. Tsukimoto, Rule extraction from neural networks via decision tree induction, in *Proceedings of the 2001 IEEE International Joint Conference on Neural Networks, IJCNN 2001, Washington, DC, USA* (2001), pp. 1870–1875

33. K.K. Sethi, D.K. Mishra, B. Mishra, Extended taxonomy of rule extraction techniques and assessment of KDRuleEx. Int. J. Comput. Appl. **50**(21), 25–31 (2012)

34. R. Srikant, R. Agrawal, Mining quantitative association rules in large relational tables, in *Proceedings of the 1996 ACM SIGMOD International Conference on Management of Data, SIGMOD'96, Montreal, Quebec, Canada* (1996)

35. G.G. Towell, J.W. Shavlik, Extracting refined rules from knowledge-based neural networks. Mach. Learn. **13**(1), 71–101 (1993)

36. H. Tsukimoto, Extracting rules from trained neural networks. IEEE Trans. Neural Netw. **11**(2), 377–389 (2000)

37. S. Ventura, J.M. Luna, *Pattern Mining with Evolutionary Algorithms* (Springer, Berlin, 2016)

38. G.I. Webb, OPUS: an efficient admissible algorithm for unordered search. J. Artif. Intell. Res. **3**, 431–465 (1995)

39. G.I. Webb, Discovering associations with numeric variables, in *Proceedings of the Seventh ACM SIGKDD International Conference on Knowledge Discovery and Data Mining (KDD-2001), New York* (2001), pp. 383–388

40. C. Zhang, S. Zhang, *Association Rule Mining: Models and Algorithms* (Springer, Berlin, 2002)

Chapter 8
Successful Applications

Abstract Supervised descriptive pattern mining has been denoted as a really important area of research, including interesting tasks such as the discovery of emerging patterns, contrast set mining, subgroup discovery, class association rule mining, exceptional models mining, among others. The aim of this chapter is to illustrate a wide range of real problems in which all these tasks have been correctly applied. In this regard, different interesting results are described in areas such as medicine, education, sociology, economy, music and traffic load prediction.

8.1 Introduction

In many application fields, there is a growing interest in data analysis, which is concerned with the development of methods and techniques for making sense of data [13]. At this point, patterns in data play a crucial role in data analysis since they represent any type of homogeneity and regularity in data and serve as good descriptors of intrinsic and important properties of data [1]. When analysing patterns, they are mainly evaluated according to the number of data records they cover so it is said that each pattern represents a data subset and its frequency is related to the size of such subset. In general, there are two types of patterns, that is, local and global patterns. Local patterns represent small parts of data, with a specific internal structure, and which distribution substantially deviate from the distribution of the whole data. Global patterns, on the contrary, represent large amounts of data so they do not usually denote an unexpected behavior. In a research study [9], it was denoted that global patterns can be easily discovered by humans and most of the provided knowledge were already discovered.

Focusing on data analysis, two main groups of tasks can be defined, that is, predictive and descriptive tasks. Predictive tasks include techniques where models, typically induced from labeled data, are used to predict a target variable (labeled variable) of previously unseen examples. On the other hand, descriptive tasks are mainly focused on finding comprehensible and interesting behavior in the form of patterns with no target variable. Even when these tasks have been researched by different communities (predictive tasks principally by the machine learning

© Springer Nature Switzerland AG 2018
S. Ventura, J. M. Luna, *Supervised Descriptive Pattern Mining*,
https://doi.org/10.1007/978-3-319-98140-6_8

community [29], and descriptive tasks mainly by the data mining community [13]), there are many situations where it is required to use both at time. In this regard, *Novak et al.* [32] described the concept of supervised descriptive pattern mining as a general task aiming at describing and understanding an underlying phenomena. This general task can be defined as a descriptive analysis in which a target variable is considered so the aim is to provide features that are highly related to this variable. Supervised descriptive discovery were originally related to three main tasks: contrast set mining [6], emerging pattern mining [12] and subgroup discovery [3]. Nowadays, however, there are many additional tasks that can be grouped under the supervised descriptive pattern mining concept (class association rules [46], exceptional models [8], discriminant discovery [39], etc).

In order to demonstrate the usefulness of any of the tasks within the supervised descriptive pattern mining area, a series of real world applications are gathered and briefly described in this chapter. Such applications have been grouped into five main groups: medicine, education, sociology, economy and energy, and others. As for the medical domain, supervised descriptive patterns have been applied to distinguish tissue samples of different forms of cancerous tumor [40]. They have been also applied to discriminate between two groups of ischaematic brain stroke patients [20] as well as to recognize coronary heart disease even before the first symptoms actually occur [10]. Supervised descriptive patterns have been also applied to identify genes that were related to colon cancer [21] and breast tumor [16], as well as to identify gene expression profiles of more than six subtypes of acute lymphoblastic leukemia patients [22].

Focusing on the educational field [36], tasks such as subgroup discovery and class association rules have been correctly considered for different purposes. The main aim of this research area is the discovery of interesting subgroups that provide the instructor with useful information about the students' learning. Additionally, subgroup discovery was considered to the study of students' attitudes in high school [31], extracting relations among different students' skills and preferences during the high school period. Even when subgroup discovery has been the main task for solving educational problems, class association rule mining has been also applied to provide feedback to instructors from multiple-choice quiz data [37].

In other real problems such as sociology, the use of supervised descriptive pattern mining techniques has provided really interesting knowledge. For example, an analysis on census data in 1991 for North West England revealed [19] that both geography and deprivation were relevant for high mortality and their interactions were also important, especially in Greater Manchester and Liverpool. Gender discrimination has also been an area of special interest for many researchers. In this regard, supervised descriptive patterns were applied to identify discrimination by gender on a dataset of scientific research proposals submitted to an Italian national call [33]. All the aforementioned real applications together with others related to economy [43], energy [11] or traffic load prediction [47] are properly described in this section.

8.2 Applications

Supervised descriptive pattern mining [32] has been considered for solving many real problems such as those related to medicine, education, sociology, and economy among others. In this section, all these real-world applications are properly described, enumerating different forms of supervised descriptive patterns that have been used to obtain useful insights.

8.2.1 Medicine

Contrast set mining has been correctly applied to the identification of patterns in synchrotron X-ray data that distinguish tissue samples of different forms of cancerous tumor [40]. Data to be analysed were gathered from patients admitted for tumor resection surgery at the Royal Melbourne Hospital, Melbourne, Australia, and the Royal University Hospital, Saskatoon, Canada. Taking the ideas from [43], authors considered a general contrast set mining algorithm discovering that glioblastomas, a type of tissue, was related to the 2nd and 4th order myelin rings. Here, four additional cases were covered than would be expected if the rings' presence was independent of the glioblastomas tumor type. According to the authors, this knowledge should be treated with caution, since the presence of myelin in samples of benign tumors was dependent on the exact location and amount of tissue excised during the surgery.

Lavrac et al. [20] proposed an approach to discriminate between two groups of ischaematic brain stroke patients, that is, those patients with thrombolic stroke and those with embolic stroke. A stroke occurs when blood supply to a part of the brain is interrupted, resulting in tissue death and loss of brain function. Embolic strokes, thrombolic strokes and stokes caused by stenosis of blood vessels are categorized as ischaemic strokes (80% of all strokes are of this type). To discriminate between two groups of ischaematic brain strokes, a dataset comprising 300 patients was analysed (data were gathered from patients that were treated at the Intensive Care Unit of the Department of Neurology, University Hospital Center Zagreb in Croatia, in year 2003). Authors focused on two main contrast sets: (*a*) embolic stroke patients were those patients with cardiovascular disorders while cardiovascular disorders were not characteristic for thrombolic stroke patients; (*b*) high age and triglyceride values, no atrial fibrillation and no anticoagulant therapy were characteristics for thrombolic stroke.

Li et al. [21] focused their studies on mining emerging patterns to analyse genes related to colon cancer. The discovered patterns denoted groups of genes that were constrained to specific intervals of gene expression levels such that patterns only occurred in one class of cells but did not occur in the other cells. It was discovered that if the expression levels of genes K03001, R76254, and D31767 in a cell were, respectively, not less than 89.20, 127.16, and 63.03, then this cell was highly probable to be a cancer cell. The experimental analysis revealed a number of jumping emerging patterns (a type of emerging patterns [12]) with a high frequency in the normal tissues and a zero frequency in cancer tissues. As an example, genes T49941, M62994, L02426, H11650, R10707 and M37721 all together presented a very large frequency (90.91%) in the normal tissues. However, no single cancer cell was satisfied for all these conditions.

In a different research work, *Li et al.* [22] considered an emerging pattern mining [7] algorithm to discover novel rules that described the gene expression profiles of more than six subtypes of acute lymphoblastic leukemia patients. Data consisted of gene expression profiles of 327 acute lymphoblastic leukemia samples and containing all the known acute lymphoblastic leukemia subtypes: T-cell (T-ALL), E2A- PBX1, TEL-AML1, MLL, BCR-ABL, and hyperdiploid (Hyperdip>50). Profiles were obtained by hydridization on the Affymetrix U95A GeneChip containing probes for 12,558 genes [45]. According to the results, very few genes could act individually as an arbitrator by itself to distinguish one subtype from others. An interesting example of emerging pattern (in the form of a rule) in this study was the following: if the expression of $38, 319\_at$ was $< 15, 975.6$, then this sample was of the type T-ALL. Additionally, if the expression of $33, 355\_at$ was $< 10, 966$, then the sample should be of the type E2A-PBX1. Many pairs of genes, however, were used in groups to separate clearly one subtype from others. For example, the experimental study denoted that if the expression of gene $40, 454\_at$ was ≥ 8280.25 and the expression of gene $41, 425\_at$ was ≥ 6821.75, then the sample was of the type E2A-PBX1 (otherwise it was one of other subtypes).

Subgroup discovery has been used in numerous real-life applications such as the analysis of coronary heart disease [10]. The aim of this type of analyses is to help general practitioners to recognize coronary heart disease even before the first symptoms actually occur. With this aim, authors considered a database with information from 238 patients that was collected at the Institute for Cardiovascular Prevention and Rehabilitation, Zagreb, Croatia. Three different groups were considered in data: (*a*) collecting anamnestic information and physical examination results, including risk factors (age, family history, weight, smoking, alcohol consumption, blood pressure, etc); (*b*) collecting results of laboratory tests (lipid profile, glucose tolerance, trombogenic factors); and (*c*) collecting measurements of heart rate, left ventricular hypertrophy, cardiac arrhytmias, etc. On this dataset, Apriori-SD [15] was considered obtaining the most representative rules for each of the three groups of data. For the first group, it was obtained that coronary heart disease was highly related to a positive family history and an age over 46 years. Additionally, it was also obtained that the disease was related to a body mass index over 25 kgm^{-2} and an age over 63 years. As for the second group, it was discovered that coronary heart disease was highly related to a total cholesterol over 6.1 $mmolL^{-1}$, an age over 52 years, and a body mass index below 30 kgm^{-2}. Additionally, in this second group, the coronary heart disease was also related to a total cholesterol over 5.6 $mmolL^{-1}$, a fibrinogen over 3.7 gL^{-1}, and a body mass index below 30 kgm^{-2}. Finally, regarding the third group, it was obtained that the coronary heart disease was related to a left ventricular hypertrophy.

Class association rules were properly applied for the task of gene expression classification [16]. In this work, authors considered microarray data containing the gene expression levels of 54,675 genes for 129 patients with breast tumor. According to the experimental results, 12 biomarker genes were identified to distinguish between HER2+ and HER2- (Human epidermal growth factor receptor2 positive or negative, respectively). PSMD3, which is a proteasome protein that regulates the ubiquitination process, had a medium small expression level in HER2- samples. GRB7, which promotes tumor growth in breast cancer cells, had a medium expression level for HER2-. Another biomarker gene is PPARBP, which had medium small expression levels to discriminate between HER2- from HER2+. As a conclusion, and based on the results of the experiments, authors considered that it was necessary a more in depth investigation of the role of these genes in breast cancer and HER2 expression.

8.2.2 Education

Subgroup discovery [14] was first applied to e-learning systems in 2009 through an evolutionary algorithm [35]. The approach considered by the authors was a modified version of one of the first evolutionary algorithms in the subgroup discovery task, that is, SDIGA (Subgroup Discovery Iterative Genetic Algorithm) [5]. The study was based on data gathered from 192 different courses at the University of Cordoba, from which only those courses (with a total of 293 students) with the highest usage of the activities and resources available in Moodle were considered. As a result, some really interesting subgroups were obtained, providing the instructor with beneficial or detrimental relationships between the use of web-based educational resources and the student's learning. As an example, it was discovered that really good marks were obtained by those students who had completed a high number of assignments and had sent a lot of messages to the forum in the ProjectManagement (C110) course. It was also discovered that most of the students who had sent a very low number of messages in the forum were highly probable to fail the exam in the AppliedComputerScienceBasis (C29) course. Through this information, the instructor might pay a higher attention to this type of students.

The subgroup discovery task has also been applied to the study of students' attitudes in high school [31]. According to the authors, the aim of this research study was to propose a system that supported the students' final degree decision by extracting relations among different students' skills and preferences during the high school period. The idea was to be able to provide advices with regard to what is the best degree option for each specific skill and student. Authors considered an evolutionary algorithm for subgroup discovery [23] and data gathered from three non-stop years at King Abdulaziz University, Saudi Arabia. A total of 5260 students were considered just in their first year of the degree in Computer Science, Engineering, Medicine, and Other. Results revealed an interesting behavior, denoting that students with the highest GPA (Grade Point Average) tended to enrol in Medicine.

Those students with a medium GPA divided their enrolment into Computer Science and Engineering. This is quite interesting since Computer Science and Engineering share similar skills, so students with a good GPA but not as good to enrol in Medicine, tended to study any of these two degrees. A different rule denoted that if a student obtained a GPA value between 4.53 and 4.97, then he/she studied the Medicine degree with a probability of 75%, and if a student obtained a GPA between 3.99 and 4.91 and his/her desire was to study Engineering, then the student studies the desired degree with a probability of 80.6%.

In a second study, authors [31] performed a different analysis based on subgroup discovery to discover the relationship between the final mark of different subjects and the degree they finally enrolled in. Here, it was discovered that the fact of obtaining a maximum mark in chemical (A+) was highly related to the fact of choosing Medicine as degree to study (87.7% of the students). This rule was denoted as a really interesting one to recommend any student to make an effort in this subject if the really wanted to study Medicine. A different relationship was obtained between a final mark of A in physics and the desire of studying Computer Science (71.9% of the students that studied Computer Science obtained an excellent final mark in physics). Finally, in a third study, authors [31] considered the use of gender as a really interesting issue, obtaining rules whose reliability changed un a huge grade when male and female students were studied.

The use of class association rules has been correctly used for educational purposes such as the analysis of students' performance on learning management systems [25]. Data under study were gathered from a Moodle system, considering a total of 230 students on 5 Moodle courses on computer science at the University of Cordoba. Information gathered was related to the course, the number of assignments done, the number of quizzes that were taken for each student, the number of passed/failed quizzes, the total time spent on quizzes, the final mark, etc. In this study, authors used an evolutionary algorithm [24] for mining rare association rules, quantifying the importance of the obtained rules through measures such as support (percentage of data records satisfied) and confidence (reliability of the rule). Results illustrated that if students spent a lot of time in the forum (a HIGH value), then they passed the final exam. According to the authors [25], this rule provided information to the instructor about how beneficial the forum was for students. Additionally, it was discovered that students of the course 110 who submitted many assignments passed the final exam, denoting therefore that the number of assignments was directly related to the final mark. Some rules included information of students who were absent for the exam, illustrating that if students did not spend time on assignments, forum participation and quizzes, then they would not take the exam. Some additional rules stated that if students did not fail any quizzes but their number of assignments was low, then they failed the final exam. This rule was very interesting because it showed how the fact of passing the quizzes was not a condition for passing the final exam.

The task of mining class association rules has been also applied to provide feedback to instructors from multiple-choice quiz data [37]. In this work, authors considered information gathered from Moodle learning management system and a multiple-choice quiz from a final exam on CLIPS programming language (a total of 104 students), at University of Cordoba. The online quiz consisted of 40 items/questions with three possible options and only one of these options was correct. Students had only one attempt at the test but they had the possibility of answering the questions in a flexible order. In a first study, the aim was to extract relationships between several questions. Here, it was obtained rules of the type: IF $Item - Num.12 = CORRECT$ AND $Item - Num.29 = CORRECT$ THEN $Item - Num.38 = CORRECT$; IF $Item - Num.24 = INCORRECT$ AND $Item - Num.35 = INCORRECT$ THEN $Item - Num.8 = INCORRECT$. In general, these two rules could show questions that could evaluate the same concepts. The instructor should check the content of these questions. In a second study, the aim was to extract rules that relate questions and the final mark. For example, the rule IF $Item - Num.1 = INCORRECT$ AND $Item - Num.29 = INCORRECT$ THEN $Grade = FAIL$ denoted that if a student failed questions 1 and 29 then he/she was highly probable to fail the exam. On the contrary, the rule IF $Item23 = CORRECT$ AND $Item31 = CORRECT$ THEN $Grade = EXCELLENT$ described that it was highly probable to obtain an excellent mark if questions 23 and 31 were correctly answered. This information was really interesting since the instructor could create a new and shorter version of the quiz by including only these items (they are the most discriminatory ones). Finally, different studies were performed by considering relationships between items, times and scores (IF $Item - Num.26 = CORRECT$ AND $Time = FEW$ THEN $Grade = GOOD$); so the instructor could consider to provide the students with higher or lower time to perform the quiz.

Class association rules have been also considered in the educational field by considering contextual-sensitive features, that is, those whose interpretation depends on some contextual information [27]. In this research work, authors considered an evolutionary algorithm for mining this type of association rules and two real problems: personalized student success factors, and graduate analysis based on Moodle learning management system. As for the personalized student success factors, a real dataset was considered, which reflected some of the aspects of the behavior of students in the Computer Science graduate program of the Technical University of Eindhoven. Four different study programs were considered: Business Information Systems, Computer Science and Engineering (CSE), Embedded Systems, and Information Security Technology. Additionally, the following undergraduate subjects (the three most common) were studied: calculus; algorithms; and datamodeling. The experimental analysis revealed that the fact of obtaining a good mark in calculus (a final mark greater or equal to 7) was related to the fact of passing the program selected by the student (reliability of 85.71%). It provided a really interesting description to be considered by students in the future. Nevertheless, it was also discovered that the feature $Program = CSE$ was a nonprimary feature since it did not provide any information about the class (pass or fail the program) and when it is combined with the previous information, it is obtained that 100% of the students that obtained a good mark in calculus (a final mark greater or equal to 7) and chose the CSE program, then they usually pass the program. According to the authors, the obtained rule was quite interesting for students that chose the CSE program, since obtaining a final mark higher or equal to 7 in the calculus subject is a prerequisite to pass the program. Finally, as for the graduate analysis based on Moodle learning management system, authors considered real data gathered from a Moodle learning management system (LMS) at the University of Cordoba, Spain. Data included 230 different students from diverse courses of the degree in Computer Science. A good description of the students' behaviors was determined by the rule IF $first\_quiz = LOW$ THEN $grade = EXCELLENT$ with a reliability of 4.16% so a small number of students that did not obtain a good mark in the first quiz obtained an excellent grade at the end of the course. In other words, it was extremely hard to obtain an excellent grade if the mark obtained in the first quiz was not satisfactory. Nevertheless, considering the contextual-feature $n\_posts = HIGH$ the reliability of the previous rule was 16.66%, so it was more probable to obtain a really good mark at the end of the course even though the mark obtained in the first quiz is low. This assertion is right if and only if the number of created posts is very high, so a student without any motivation at the beginning, could improve his/her motivation during the course and could obtain a very good mark at the end of the course.

8.2.3 Sociology

Subgroup discovery [14] has been correctly applied to the census data [19]. Here, the census data in 1991 for North West England, one of the twelve regions in UK, was considered. The geographical units used in the analyses were the 1011 districts (an average of 7000 habitants were considered for each district) situated in the 43 local authorities of North West England. In this problem, authors applied SubgroupMiner [18] to discover geographical factors for high mortality. A map of mortality (the target variable) illustrated a clear geographical distribution with areas of high mortality in Manchester and Liverpool and other urban areas. Obtained results revealed that 85% of the 33 districts in Manchester were high mortality districts, which was significantly higher than the 25% of high mortality districts in North West England. Then, authors considered the use of deprivation indices as explanatory variables, which were generally used for socio-economic studies. The results supported the assumption that both geography and deprivation were relevant for high mortality and their interactions were also important, especially in Greater Manchester and Liverpool. Finally, authors in [19] considered the system Explora [17] and they selected lung cancer mortality as the target variable. Some results for this particular mortality rate which deviated from the results for the overall mortality rate were also summarized.

Gender discrimination has been an area of special interest for many researchers. An example of that is the study that [33] was carried out on a dataset of scientific research proposals submitted to an Italian national call. In this research work, authors considered rules including the attribute $gender = female$ in the antecedent of the rule. In this regard, the first rule highlighted a context for a specific research area, that is, life sciences. This rule covered a considerable number of data records (proposals) among those labeled as discriminated, that is, 9.5% of the total (27% of those in the life sciences research area), with a precision of 100%. The rule denoted research proposals requiring two or more young researchers, having a cost for them of more than 244,000 EUR or more, and it was required that the principal investigator had at most 12 publications with a mean number of authors of 8.4 or more. According to the authors, the lack of knowledge about the skills of an individual could be compensated by the peer-reviewers of life sciences through a prior knowledge of the average performances of the group or category the individual belongs to, in our case the gender. Additionally, a large context of possible discrimination was highlighted by another rule that concentrated on proposals with high budget led by young principal investigators. According to the authors, this could be denoted as if peer-reviewers of panel *Physical and Analytical Chemical Sciences* trusted young females requiring high budgets less than males leading similar projects.

In 2013 *Romei et al.* [34] presented a case study on gender discrimination in a dataset of scientific research proposals, distilling from the case study a general discrimination discovery process. Data referred to two research areas for an Italian call for research proposals with 3790 applications nationwide. According to the authors, the aim of this research study was to analyse whether there are different success rates of males and females due to legitimate characteristics or skill differences of the gender of applicants. Analysing the dataset, it was observed that proposals led by females required slightly lower costs for young researchers than proposals led by males. Analysing the total cost and the requested grant the situation was quite similar since the average total cost was 980K EUR for females and 1080K EUR for males. Results on the dataset discovered some interesting behaviour concerning proposals with high budget for young researchers and for good reputation researchers. There were 16 proposals led by male researchers, four of which passed the peer-review, and 14 proposals led by female researchers, none of which passed it. Concerning young researchers with a fund request greater or equal than 310K EUR, there were 201 proposals, 131 led by male researchers (16 of which passed the peer-review) and 70 led by female researchers (only one of which passed the peer-review).

8.2.4 Economy and Energy

Contrast set mining has been applied to areas such as economy, analyzing retail sales data [43]. The analysis on this field was based on contrasting the pattern of retail activity on two different days with the aim of highlighting the effect of specific marketing promotions on purchasing behavior. The STUCCO [4] algorithm was used to perform the experimental analysis, revealing that the proportion of customers buying from each of departments 851 and 855 on the second day was higher than the first. This effect was heightened when customers that bought from both departments in a single transaction were considered. Additionally, it was obtained that whereas items for departments 220 and 355 were each purchased more frequently on August 14th than August 21st, a greater proportion of customers bought items from both departments on the 21st than the 14th.

Another application example in the area of economy was described in [44], considering contrast set mining with the aim of finding meaningful sets that can assist in the design of insurance programs for various types of customers (categorized as types of families). In this study, authors considered a dataset comprising 9852 records and divided into 10 groups according to the type of customer: successful hedonists, driven growers, average family, career loners, living well, cruising seniors, retired and religious, family with grown ups, conservative families, and farmers. From a total of 85 attributes, authors considered just those 20 attributes that were related to policies for insurance programs. The obtained results when running the STUCCO [4] algorithm demonstrated some useful information about relationships among the insurance policies for various types of families. For most of the families, their insurance program did not include contribution clauses to cars, motorbikes and fire at the same time. However, there was an exception for those families categorized as successful hedonists and living well, or maybe cruising seniors. Most families did not have contribution car policies and contribution moped policies except those people who had enough spare time to enjoy their livings. According to the findings [44] it was surprising that most families did not choose contribution fire policies together with either contribution car policies or contribution moped policies. Additionally, conservative families were the only group that did not consider contribution life policies when either contribution car policies or contribution fire policies were already considered. It was also discovered that, even when contribution private third party insurance for agriculture and contribution tractor policies were primarily designed for farmers, this group of family is unlikely to have them both. Results also revealed that farmers were not interested in owning contribution fire policies when their insurance program included contribution private third party insurance for agriculture, contribution car and contribution tractor policies.

The price analysis of rental residences of apartment buildings has been considered by different researchers [6]. According to authors, it is a general thought that the value of houses is calculated according to attributes such as the occupied area, the distance from the nearest station, the building age, and facilities among others [38]. However traditional analysis has not considered the information of the room layout through the evaluation of the usability and fineness of space. The aim was to select structures that were highly related to rent price (target variable) by examining emerging patterns. Results obtained from the experimental analysis revealed some interesting distributions that were really important in the value of the house: the segregation level of a dining from an entrance and a hall; the presence or the absence of a separate kitchen; and the position of a Japanese style room.

Song et al. [41] proposed a methodology to detect changes in the shopping habits automatically from the customers' profiles and sales data at different time snapshots. In order to discover unexpected changes, the emerging pattern mining task was applied on data gathered from a Korean online shopping mall. This dataset, which was divided into two subsets according to different periods of time, contains customer profiles and purchasing information including age, job, sex, address, number of purchases, number of visits, and payment method among others. The experimental stage revealed a series of patterns with a high growth (more than 90%) in sales for customers who were specialists and visit the mall frequently so a marketing campaign to invoke the revisiting should be developed. Additionally, it was also obtained a growth in sales for customers who lived in different cities and visited the mall frequently. It was also found that sales for female customers who lived in KyungNam were low from the first dataset. However, when considering the second dataset, then sales for female customers who visited the mall frequently were high even if they were living in KyungNam. According to the authors it denoted that the importance of customers who lived in KyungNam and visited the mall frequently was gradually increasing. Therefore, a modification for the existing marketing strategy and plan was required.

In 2016, *Garcia et al.* [11] proposed the use of emerging patterns to discover interesting behaviour on a Concentrating Photovoltaic (CPV) problem which is an alternative to the conventional photovoltaic for the electric generation. Data for the experimental analysis were gathered from a CPV module with 25 solar cells and a concentration factor of 550. The measures were acquired at the rooftop of a building at the University of Jaen during the period between March 2013 and November 2013. According to the results, the best growth ratios were obtained when the direct normal irradiance was low and a low power. Another really good emerging pattern was obtained for the features: low average photon energy, low ambient temperature, medium wind speed, high spectral machine radio and high incident global irradiance. As a general knowledge, when combining all the extracted patterns, it was obtained that when power is maximum the value for the average photon energy is low. In this way and according to the authors, there was an interesting further analysis with respect to the influence of this variable in the performance of the CPV module.

Subgroup discovery was correctly applied to marketing [5] by proposing a completely new algorithm, called SDIGA (Subgroup Discovery Iterative Genetic Algorithm), which has been properly described in Chap. 4. According to the authors in [5], and focusing on the area of marketing (planning of trade fairs), it is of high interest to extract conclusions from the information on previous trade fairs to determine the relationship between the trade fair planning variables and the success of the stand. Data gathered from the Machinery and Tools biennial fair held in Bilbao, Spain, in March 2002, were considered. The dataset contained information on 228 exhibitors, and for each exhibitor the stands were characterized according to their level of achievement of objectives (low, medium, or high efficiency). The experimental study carried out revealed that the trade fair was held in the North zone, and the exhibitors were coming principally from that zone, so worse results for the exhibitors coming from more distant zones were explained due to the distance and the low level of knowledge of the peculiarities of the exhibition. Additionally, the exhibitors who obtained best results were those from the Central zone and who did not send a thank-you pamphlet to all the contacts. They were generally large (or medium-sized companies) with either a very high or a low annual sales volume. These results demonstrated that big companies were able to invest lots of money preparing the trade fair, so the results they were better. On the contrary, small companies spent little money on the trade show, and therefore their expectations about their participation were poor.

Exceptional model mining was considered in the analysis of the housing price per square meter, as it was demonstrated by *Duivesteijn et al.* [8]. Authors considered the Windsor housing dataset [2] comprising information on 546 houses that were sold in Windsor, Canada in the summer of 1987. The experimental analysis revealed that the following subgroup (and its complement) was found to show the most significant difference in correlation: $drive = 1 \wedge rec\_room = 1 \wedge nbath \geq 2.0$. This subgroup represented a set of 35 houses comprising a driveway, a recreation room and at least two bathrooms. The correlation for this subgroup was -0.090, whereas set of the remaining 511 houses obtained a correlation of 0.549. According to the authors in [8], a tentative interpretation could be that the subgroup described a collection of houses in the higher segments of the markets where the price of a house was mostly determined by its location and facilities. The desirable location could provide a natural limit on the lot size, such that this was not a factor in the pricing.

8.2.5 Other Problems

The extraction of class association rules have been applied to some additional problems such as the one of traffic load prediction [47]. To demonstrate that class association rules could be useful for traffic prediction, authors considered a simulator and a 7×7 road map model, that is, each section had the same length and all cars had the same timer but they did not exactly have the same speed (it depends on the concrete traffic situations). All the cars used the optimal route by Q Value-based Dynamic Programming [28]. Additionally, each section between two intersections in the road had two directions, and each of these directions was represented by a series of attributes. From the set of discovered rules, an example is the following: IF $W4N5, W4N6, Low(t = 0)$ AND $W4N5, W3N5, High(t = 9)$ THEN $W3N5, W3N6, High(t = 11)$. This rule determined that if the sections on the map named $W4N5, W4N6$ and $W4N5, W3N54$ had low traffic at time 0 and 9, respectively, then, the section named $W3N5, W3N64$ would have high traffic volume at time 11. As a result, the proposed method was able to extract important time-related class association rules that were of high interest to improve the traffic flow.

Exceptional association rules had been applied to the analysing unusual and unexpected relationships between quality measures in association rule mining [26]. In the experimental evaluation, authors conducted a case study on mining exceptional relations between well-known and widely used quality measures of association rules [1, 42], that is, Support, Confidence, Lift, Leverage, Conviction, CF and IS. The study was carried out on a dataset including seven attributes (one per quality measure) and 348,511 records, that is, combinations of values for all the quality measures. As a result, it was obtained that despite the fact that Support and Leverage seemed to be positively correlated, they were negatively correlated when considering a value in the range $[0.2, 0.4]$ for the IS measure and $[-0.7, 0.5]$ for the CF measure. Additionally, it was obtained that even when Support and CF were positively correlated, the fact of considering IS $\in [0.5, 0.9]$ and Lift $\in [0.1, 41.0]$ made the correlation negative and, therefore, completely different.

In a recent study, *Neubarth et al.* [30] proposed the use of contrast set mining for the discovery of recurrent patterns in groups of songs of Native American music. The analysis was performed on a set of more than 2000 songs Native American tribes collected and transcribed over seven decades, from 1893 to the late 1950s. A series of features were considered in data including tonality (major, minor, third lacking, irregular, etc), firstReKey (keynote, triad tone within octave, other tone within octave, etc), compass (4 or less, 5–8, 9–12, more than 12), among others. Among other interesting patterns, the one $\{initMetre : duple, metreChange : no\}$ was particularly supported by Choctaw songs. Here, double time was preferred by the Choctaw for the beginning of their songs. In terms of metre changes, Choctaw songs showed the smallest percentage in the songs under analysis. The supervised descriptive analysis revealed therefore that the combination of these two features was almost threefold over-represented in the Eastern area compared to other areas. Similarly, songs with the feature $\{metreChange : no\}$ were found more frequently in the California-Yuman style than in other areas. It was also obtained that 59% of California-Yuman songs satisfied the pattern $\{compass : 5 \, to \, 8, lastReCompass : containing\_lower\}$, against only 23% of songs in other areas.

References

1. C.C. Aggarwal, J. Han, *Frequent Pattern Mining* (Springer International Publishing, Cham, 2014)
2. P. Anglin, R. Gencay, Semiparametric estimation of a hedonic price function. J. Appl. Economet. **11**(6), 633–48 (1996)
3. M. Atzmueller, Subgroup discovery - advanced review. WIREs Data Min. Knowl. Discov. **5**, 35–49 (2015)
4. S.D. Bay, M.J. Pazzani, Detecting group differences: mining contrast sets. Data Min. Knowl. Discov. **5**(3), 213–246 (2001)
5. M.J. del Jesus, P. Gonzalez, F. Herrera, M. Mesonero, Evolutionary fuzzy rule induction process for subgroup discovery: a case study in marketing. IEEE Trans. Fuzzy Syst. **15**(4), 578–592, (2007)
6. G. Dong, J. Bailey (eds.), *Contrast Data Mining: Concepts, Algorithms, and Applications* (CRC Press, Boca Raton, 2013)
7. G. Dong, J. Li, Efficient mining of emerging patterns: discovering trends and differences, in *Proceedings of the 5th ACM SIGKDD International Conference on Knowledge Discovery and Data Mining (KDD-99)*, New York, 1999, pp. 43–52

8. W. Duivesteijn, A. Feelders, A.J. Knobbe, Exceptional model mining - supervised descriptive local pattern mining with complex target concepts. Data Min. Knowl. Discov. **30**(1), 47–98 (2016)

9. J. Fürnkranz, From local to global patterns: evaluation issues in rule learning algorithms, in *International Seminar on Local Pattern Detection*, Dagstuhl Castle, 2004, pp. 20–38

10. D. Gamberger, N. Lavrac, Expert-guided subgroup discovery: methodology and application. J. Artif. Intell. Res. **17**(1), 501–527 (2002)

11. A.M. García-Vico, J. Montes, J. Aguilera, C.J. Carmona, M.J. del Jesus, Analysing concentrating photovoltaics technology through the use of emerging pattern mining, in *Proceedings of the 11th International Conference on Soft Computing Models in Industrial and Environmental Applications (SOCO 16)*, San Sebastian, October 2016, pp. 334–344

12. A.M. García-Vico, C.J. Carmona, D. Martín, M. García-Borroto, M.J. del Jesus, An overview of emerging pattern mining in supervised descriptive rule discovery: taxonomy, empirical study, trends and prospects. WIREs Data Min. Knowl. Discov. **8**(1), e1231 (2017)

13. J. Han, M. Kamber, *Data Mining: Concepts and Techniques* (Morgan Kaufmann, Amsterdam, 2000)

14. F. Herrera, C.J. Carmona, P. González, M.J. del Jesus, An overview on subgroup discovery: foundations and applications. Knowl. Inf. Syst. **29**(3), 495–525 (2011)

15. B. Kavšek, N. Lavrač, APRIORI-SD: adapting association rule learning to subgroup discovery. Appl. Artif. Intell. **20**(7), 543–583 (2006)

16. K. Kianmehr, M. Kaya, A.M. ElSheikh, J. Jida, R. Alhajj, Fuzzy association rule mining framework and its application to effective fuzzy associative classification. Wiley Interdiscip. Rev. Data Min. Knowl. Discov. **1**(6), 477–495 (2011)

17. W. Klösgen, Explora: a multipattern and multistrategy discovery assistant, in *Advances in Knowledge Discovery and Data Mining*, ed. by U.M. Fayyad, G. Piatetsky-Shapiro, P. Smyth, R. Uthurusamy (American Association for Artificial Intelligence, Menlo Park, 1996), pp. 249–271

18. W. Kloesgen, M. May, Census data mining - an application, in *Proceedings of the 6th European Conference on Principles and Practice of Knowledge Discovery in Databases*, PKDD 2002, Helsinki (Springer, London, 2002), pp. 733–739

19. W. Klosgen, M. May, J. Petch, Mining census data for spatial effects on mortality. Intell. Data Anal. **7**(6), 521–540 (2003)

20. N. Lavrac, P. Kralj, D. Gamberger, A. Krstacic, Supporting factors to improve the explanatory potential of contrast set mining: analyzing brain ischaemia data, in *Proceedings of the 11th Mediterranean Conference on Medical and Biological Engineering and Computing (MEDICON-07)*, Ljubljana, June 2007, pp. 157–161

21. J. Li, L. Wong, Identifying good diagnostic gene groups from gene expression profiles using the concept of emerging patterns. Bioinformatics **18**(10), 1406–1407 (2002)

22. J. Li, H. Liu, J.R. Downing, A.E. Yeoh, L. Wong, Simple rules underlying gene expression profiles of more than six subtypes of acute lymphoblastic leukemia (ALL) patients. Bioinformatics **19**(1), 71–78 (2003)

23. J.M. Luna, J.R. Romero, C. Romero, S. Ventura, On the use of genetic programming for mining comprehensible rules in subgroup discovery. IEEE Trans. Cybern. **44**(12), 2329–2341 (2014)

24. J.M. Luna, J.R. Romero, S. Ventura, On the adaptability of G3PARM to the extraction of rare association rules. Knowl. Inf. Syst. **38**(2), 391–418 (2014)

25. J.M. Luna, C. Romero, J.R. Romero, S. Ventura, An evolutionary algorithm for the discovery of rare class association rules in learning management systems. Appl. Intell. **42**(3), 501–513 (2015)

26. J.M. Luna, M. Pechenizkiy, S. Ventura, Mining exceptional relationships with grammar-guided genetic programming. Knowl. Inf. Syst. **47**(3), 571–594 (2016)

27. J.M. Luna, M. Pechenizkiy, M.J. del Jesus, S. Ventura, Mining context-aware association rules using grammar-based genetic programming. IEEE Trans. Cybern. 1–15 (2018). Online first. https://doi.org/10.1109/TCYB.2017.2750919

28. M.K. Mainali, K. Shimada, S. Mabu, S. Hirasawa, Optimal route of road networks by dynamic programming, in *Proceedings of the 2008 International Joint Conference on Neural Networks, IJCNN*, Hong Kong, 2008, pp. 3416–3420

29. T.M. Mitchell, *Machine learning*. McGraw Hill Series in Computer Science (McGraw-Hill, New York, 1997)

30. K. Neubarth, D. Shanahan, D. Conklin, Supervised descriptive pattern discovery in native American music. J. New Music Res. **47**(1), 1–16 (2018). Online first. https://doi.org/10.1080/09298215.2017.1353637

31. A.Y. Noaman, J.M. Luna, A.H.M. Ragab, S. Ventura, Recommending degree studies according to students' attitudes in high school by means of subgroup discovery. Int. J. Comput. Intell. Syst. **9**(6), 1101–1117 (2016)

32. P.K. Novak, N. Lavrač, G.I. Webb, Supervised descriptive rule discovery: a unifying survey of contrast set, emerging pattern and subgroup mining. J. Mach. Learn. Res. **10**, 377–403 (2009)

33. A. Romei, S. Ruggieri, F. Turini, Discovering gender discrimination in project funding, in *Proceedings of the 12th IEEE International Conference on Data Mining Workshops, ICDM*, Brussels, December 2012, pp. 394–401

34. A. Romei, S. Ruggieri, F. Turini, Discrimination discovery in scientific project evaluation: a case study. Expert Syst. Appl. **40**(15), 6064–6079 (2013)

35. C. Romero, P. González, S. Ventura, M.J. del Jesús, F. Herrera, Evolutionary algorithms for subgroup discovery in e-learning: a practical application using Moodle data. Expert Syst. Appl. **36**(2), 1632–1644 (2009)

36. C. Romero, S. Ventura, M. Pechenizkiy, R.S.J. Baker, *Handbook of Educational Data Mining*. CRC Data Mining and Knowledge Discovery Series (Chapman and Hall, CRC Press, 2011)

37. C. Romero, A. Zafra, J.M. Luna, S. Ventura, Association rule mining using genetic programming to provide feedback to instructors from multiple-choice quiz data. Expert Syst. **30**(2), 162–172 (2013)

38. R.A. Rubin, Predicting house prices using multiple listings data. J. Real Estate Finance Econ. **17**(1), 35–59 (1998)

39. S. Ruggieri, D. Pedreschi, F. Turini, Data mining for discrimination discovery. ACM Trans. Knowl. Discov. Data **4**(2), 1–40 (2010)

40. K.K.W. Siu, S.M. Butler, T. Beveridge, J.E. Gillam, C.J. Hall, A.H. Kaye, R.A. Lewis, K. Mannan, G. McLoughlin, S. Pearson, A.R. Round, E. Schultke, G.I. Webb, S.J. Wilkinson, Identifying markers of pathology in SAXS data of malignant tissues of the brain. Nucl. Instrum. Methods Phys. Res. A **548**, 140–146 (2005)

41. H.S. Song, J.K. Kimb, H.K. Soung, Mining the change of customer behavior in an internet shopping mall. Expert Syst. Appl. **21**(3), 157–168 (2001)

42. S. Ventura, J.M. Luna, *Pattern Mining with Evolutionary Algorithms* (Springer, Berlin, 2016)

43. G.I. Webb, S.M. Butler, D.A. Newlands, On detecting differences between groups, in *Proceedings of the 9th ACM SIGKDD International Conference on Knowledge Discovery and Data Mining*, Washington, DC, 24 August 2003, pp. 256–265

44. T.T. Wong, K.L. Tseng, Mining negative contrast sets from data with discrete attributes. Expert Syst. Appl. **29**(2), 401–407 (2005)

45. E.J. Yeoh, M.E. Ross, S.A. Shurtleff, W.K. Williams, D. Patel, R. Mahfouz, F.G. Behm, S.C. Raimondi, M.V. Relling, A. Patel, C. Cheng, D. Campana, D. Wilkins, X. Zhou, J. Li, H. Liu, C.H. Pui, W.E. Evans, C. Naeve, L. Wong, J.R. Downing, Classification, subtype discovery, and prediction of outcome in pediatric acute lymphoblastic leukemia by gene expression profiling. Cancer Cell **1**(2), 133–143 (2002)

46. C. Zhang, S. Zhang, *Association Rule Mining: Models and Algorithms* (Springer, Berlin, 2002)

47. W. Zhou, H. Wei, M.K. Mainali, K. Shimada, S. Mabu, K. Hirasawa, Class association rules mining with time series and its application to traffic load prediction, in *Proceedings of the 47th Annual Conference of the Society of Instrument and Control Engineers (SICE 2008)*, Tokyo, August 2008, pp. 1187–1192

Printed in the United States
By Bookmasters